SpringerBriefs in Applied Sciences and Technology

SpringerBriefs present concise summaries of cutting-edge research and practical applications across a wide spectrum of fields. Featuring compact volumes of 50 to 125 pages, the series covers a range of content from professional to academic.

Typical publications can be:

- A timely report of state-of-the art methods
- An introduction to or a manual for the application of mathematical or computer techniques
- A bridge between new research results, as published in journal articles
- A snapshot of a hot or emerging topic
- An in-depth case study
- A presentation of core concepts that students must understand in order to make independent contributions

SpringerBriefs are characterized by fast, global electronic dissemination, standard publishing contracts, standardized manuscript preparation and formatting guidelines, and expedited production schedules.

On the one hand, **SpringerBriefs in Applied Sciences and Technology** are devoted to the publication of fundamentals and applications within the different classical engineering disciplines as well as in interdisciplinary fields that recently emerged between these areas. On the other hand, as the boundary separating fundamental research and applied technology is more and more dissolving, this series is particularly open to trans-disciplinary topics between fundamental science and engineering.

Indexed by EI-Compendex, SCOPUS and Springerlink.

More information about this series at https://link.springer.com/bookseries/8884

Luciana da Costa Carvalho

Beyond Copper Soaps

Characterization of Copper Corrosion
Containing Organics

 Springer

Luciana da Costa Carvalho
School of Archaeology
University of Oxford
Oxford, UK

ISSN 2191-530X   ISSN 2191-5318 (electronic)
SpringerBriefs in Applied Sciences and Technology
ISBN 978-3-030-97891-4   ISBN 978-3-030-97892-1 (eBook)
https://doi.org/10.1007/978-3-030-97892-1

This Springer imprint is published by the registered company Springer Nature Switzerland AG
The registered company address is: Gewerbestrasse 11, 6330 Cham, Switzerland

To my mother, who introduced me to science, and to my godson, Matthew Collins, whom I try to inspire.

Foreword

I believe the work presented in this book (and the DPhil thesis upon which it is based) represents a major step forward in two different aspects of the scientific investigation of archaeological metals. Firstly, it provides a significant extension of the study of organic residues in archaeology. It has been known since the 1970s that ceramics can harbour organic residues, which provide crucial evidence about the use of these vessels, specifically about dietary and culinary practices in the past. These residues can be visible, burnt food deposits on the surface of cooking vessels, but more interestingly they can also be invisible, absorbed into the pore structure of the vessel, which protects the residue from organic degradation in a way which is yet to be fully understood. Part of the impetus for the original thesis was to extend this study to the extraction of residues from metal objects—do the corrosion products forming on metal surfaces provide a protective environment for endogenous organic molecules, particularly given the bactericidal properties of some metals, including copper? The answer is a resounding "yes", which, at a stroke, immediately extends the potential for studying organic materials in the past.

Secondly, and perhaps even more fundamentally, it demands an extension of the routine methodology for studying metal corrosion products. As noted in the introduction, it has been long understood in the field of fine art conservation that some metal-organic compounds can form on paintings as a result of the interaction between metal ions and the organic compounds in varnishes or from interactions with gases in the atmosphere. However, with a few notable exceptions, the possibilities of metal-organic complexes contributing to the formation of metal patina or corrosion products have largely been ignored. With hindsight, this is largely a consequence of the analytical approach taken, primarily the use of X-ray diffraction, which by definition, only identifies highly crystalline materials, which are largely inorganic mineral species. The chemistry becomes more complicated when we contemplate metal-organic complexes as a significant component of corrosion products, requiring advanced methods of mass spectrometry to identify these species.

It is a great tribute to Luciana's vision and drive that she managed to demonstrate the significance of these two aspects of materials science applied to archaeological metals. I expect to see some significant developments in these areas over the next few years.

A. M. Pollard
Edward Hall Professor
of Archaeological Science
University of Oxford
Oxford, UK

Preface

This book represents part of a PhD research conducted at the School of Archaeology (University of Oxford) aimed at demonstrating archaeological metal corrosion's potential for preserving organic residues [1]. Organic residue analysis is an established field in archaeological science. However, the identification of residues is complicated by chemical changes due to the processing of substances (e.g. mixing and cooking) in the past, their degradation during deposition and modern contamination. Notwithstanding these complexities, an increasing number of chemical markers for biomaterials are being reported in the literature, mostly recovered from porous ceramics. Metal corrosion has seldom been considered a locus for organic residue preservation in spite of the range of macro-organic residues found preserved within corrosion products.

In a pilot study of "unusual" bronze corrosion, [2] researchers recovered molecular markers for pine resin and oil from a *patera* (a pan-like object found in ritual contexts throughout the Roman empire). It was suggested that the methods used would only have recovered organic residues encapsulated by corrosion products, not metal-organic or organometallic complexes.

Although the study of metal-organic complexes, particularly metal carboxylates, is a popular topic amongst art conservators, their identification in archaeological contexts has seldom been reported. Archaeological metal corrosion has traditionally been considered a mixture of inorganic compounds, so its characterization is often restricted to analytical techniques able to detect these compounds. Because metals can react spontaneously with organic compounds to form metal-organic complexes, I decided to investigate metal-organic complexation as a pathway for organic residue preservation in metal corrosion, focusing on copper.

Given the difficulty in corroding copper in the soil, I set out to explore electrochemistry methods to produce corrosion products containing inorganic and organic phases. These corrosion products "impregnated with organics" would then be used to evaluate analytical techniques. However to do this I needed apparatus and access to analytical techniques beyond what was available in my department.

The first connection I established was with the Department of Chemistry, through Dr. Phil Wiseman. At the time, Dr. Wiseman controlled access to a powder X-ray Diffractometer (XRD) available, by special arrangement, to other research groups within the university. XRD is the most popular technique for the identification of mineral phases in metal corrosion so access to it was crucial.

During my first visit to Dr. Wisemen's office I got lost, ending up in a teaching laboratory. I took the opportunity to enquire about a power supply (needed for the electrochemical experiments) and the technician suggested I speak with Richard. He had meant Prof. Richard Compton, a world-renowned specialist in fundamental and applied electrochemistry!

I emailed Prof. Compton and we met the following week to discuss the feasibility of my ideas. Prof. Compton then directed me to Begbroke Science Park (part of the Materials Science Department) for the necessary apparatus —just mention my name, he said.

Without knowing who to approach at Begbroke, I called reception. As the receptionist tried to establish the best person to assist me, Dr. Colin Johnston (who apparently happened to be walking through the room) overheard the conversation and offered to help. A week later I started my experiments assisted by Mr. Chris Salter, who supervises research in archaeometallurgy.

At the time, I was trying to follow a protocol for creating copper corrosion products using electrochemistry I had found in the literature. The first step of this protocol required heating a copper coupon in a tube furnace to create a layer of copper(I) oxide. Incidentally, the tube furnace at Begbroke had belonged to archaeology!

At Begbroke, I was able to confirm the state of oxidation of copper ions in my copper-organic complexes using X-ray Photoelectron Spectroscopy (XPS) and investigate their level of crystallinity with a more powerful XRD equipment than the one at Chemistry. Moreover, I also used their Fourier Transform Infrared (FTIR) for analysis of samples using the KBr method. In the end, Dr. Johnston became my supervisor and Mr. Salter one of the assessors of the work presented in this book.

My quest for analytical equipment also led me to Mr. David Howell, former heritage scientist at the Bodleian Library. Mr. Howell had a FTIR with an ATR microscope attachment and a Raman spectroscopy setup for pigment identification in manuscripts. As I was not able to generate useful data from my copper soaps with his Raman's laser, Mr. Howell introduced me to Prof. Tony Parker, who was based at the Central Laser Facility at the Science and Technology Facilities Council. I have fond memories of the time spent with Prof. Parker testing various types of lasers on my copper soaps.

Mr. Howell invited me to present the preliminary results on copper soaps as part of his talk about research conducted at the Bodleian Library at the 2017 Perkin Elmer's FTIR Workshop. Preparing my slides enabled me to reconnect with my research which had been put on hold due to the sudden death of my father. At the workshop I met an eclectic cohort of scientists including Dr. Chris Dyer, a Raman specialist. Dr. Dyer was very enthusiastic about my research. He invited me to his laboratory at Cranfield University where we spent enjoyable days analysing the copper-organic complexes.

A substantial element of my research was the incorporation of mass spectrometry techniques in the characterization of corrosion products. As I wasn't making much progress with the Gas Chromatography with Mass Spectrometry (GC-MS) equipment available at the School of Archaeology I approached Prof. James McCullagh, Head of the Mass Spectrometry Facility at the Department of Chemistry for a collaboration. Prof. McCullagh put me under the wings of Dr. James Wickens (the laboratory manager at the time) who introduced me to the Thermal Separation Probe (TSP). Back in 2016, the TSP was a relatively new way to volatilize a sample for GC-MS analysis.

In 2019, as I was close to completing the experimental part of my PhD, I started a collaboration with the Ashmolean Museum and was offered some valuable corrosion samples from bronze objects from Pompeii, which had arrived at the museum for conservation in preparation for a special exhibition. Around the same time, I had also started a collaboration with Mrs. Elisabete Pires, the McCullagh Group's proteomics specialist.

Mrs. Pires developed the bottom-up proteomics protocol used for the characterization of Cu-Casein and Cu-Milk corrosion presented in this book. Amongst the archaeological samples analysed by Mrs. Pires was a sample of the red cinnabar paint from a 1,000-year-old mask from Peru where human blood and egg proteins were identified. The extensive press coverage our article [3] received, the only data we have published to date, exemplifies the power of interdisciplinary collaborations for the field of cultural heritage.

Oxford, UK   Luciana da Costa Carvalho

References

1. Carvalho LdC (2021) Recovery and identification of organic residues from metal corrosion. DPhil Thesis, University of Oxford
2. Merriman KRP, Ditchfield D, Goodburn-Brown D, et al (2017) Where bio- and geochemistry meet: Organic Residues in Copper-corrosion Products? In: Kluiving S, Kootket L, Hermans R (eds) Interdisciplinarity between Humanities and Science: A fetschrift in honour of Prof. Dr. Henk Kars. Sidestone Press, Leiden, pp. 177–183
3. Pires E, Carvalho LC, Shimada I, et al (2021) Human Blood and Bird Egg Proteins Identified in Red Paint Covering a 1000-Year-Old Gold Mask from Peru. J Proteome Res 20(11):5212 5217

Acknowledgements

Petros London Limited for financing my studies and Springer Nature for supporting the publication of this book.

Prof. Mark Pollard, Dr. Colin Johnston and Prof. James McCullagh for their supervision. Mr. Chris Salter, Dr. James Wickens, Dr. Phil Holdway, Dr. Chris Dyer and Mrs. Elisabete Pires for their analytical support.

Dr. John Merkel and Dr. Juliano Chaker for reviewing earlier versions of this manuscript.

About This Book

Metal corrosion has traditionally been considered a mixture of inorganic compounds. This book reports a series of electrochemical experiments where copper was corroded in the presence of various organic substances. Combining data from spectroscopy techniques, X-ray diffraction and mass spectrometry (including proteomics), the experiments demonstrate that copper-organic complexes can be formed during the corrosion of copper. The low solubility of copper-organic complexes in organic solvents and their amorphous nature mean that these compounds cannot be easily detected by one single analytical technique. This book will benefit researchers investigating the presence of organic residues in archaeological copper corrosion and copper-organic complexes in art, where sampling is often subject to curatorial constraints.

Contents

Abbreviations

ATR	Attenuated Total Reflectance
BE	Binding Energy
BSTFA	N,O-Bis(trimethylsilyl)trifluoroacetamide
CI	Chemical Ionization
COD	Crystallography Open Database
EI	Electron Ionization
EPR	Electron Paramagnetic Resonance
FDR	False Discovery Rate
FIA	Flow Injection Analysis
FTIR	Fourier Transform Infrared Spectroscopy
GC-MS	Gas Chromatography with Mass Spectrometry
NMR	Nuclear Magnetic Resonance
PSM	Peptide Spectrum Match
QTOF	Quadrupole Time-of-Flight
TSP	Thermal Separation Probe
XPS	X-ray Photoelectron Spectroscopy
XRD	X-ray Diffraction

Chapter 1
Copper-Organic Complexes in Cultural Heritage

Copper is one of the earliest metals known to man. Although found naturally in its metallic state, copper exists mainly combined with other elements as mineral deposits. Copper belongs to the *d-block* of the Periodic Table, where elements in their metallic state are usually lustrous, ductile and good conductors of heat and electricity [1]. As a *transition metal*, copper ions can have different states of oxidation (varying from +1 to +3) and form coordination compounds.

In broad terms, coordination compounds (or complexes) are formed when a Lewis base[1] (ligand) is attached to a Lewis acid[2] (acceptor) via a lone pair of electrons from a donor atom (such as oxygen and nitrogen) in the ligand. The nature of the bond between transition metals and ligands in these complexes is difficult to define and many theories exist to describe them [2]. Copper ions interact with donor atoms in various types of organic molecules, including carboxylic acids, carbohydrates and proteins, resulting in copper-organic complexes that lack a direct copper-carbon interaction.

In cultural heritage contexts, copper-organic complexes are sometimes identified as synthetic pigments. They are also identified during the conservation of paintings and copper alloy objects. The most widely known copper-organic complex pigment is *verdigris* [3, 4]. It is produced from the exposure of metallic copper to acidic vapours, resulting in a corrosion product that contains basic or neutral copper acetates. Its colour and precise chemical composition vary according to the recipe. Some recipes for verdigris include metal pre-treatment with solid remains of grapes from wine production, urine and honey [3–5], thus leading to the formation of other types of copper-organic compounds.

Verdigris is also used as an ingredient to produce other synthetic pigments [3, 5]. When combined with resins rich in abietic acid, it forms copper resinates, a mixture of copper carboxylates including those where the acetate ligand has been replaced

---

[1] Any substance or ion that can donate a pair of electrons.

[2] Any substance or ion that can accept a pair of electrons.

© The Author(s), under exclusive license to Springer Nature Switzerland AG 2022
L. C. Carvalho, *Beyond Copper Soaps*,
SpringerBriefs in Applied Sciences and Technology,
https://doi.org/10.1007/978-3-030-97892-1_1

by abietate [3, 6–8]. Verdigris mixed with a proteinaceous substance (e.g. egg yolk) forms transparent copper proteinates used to illuminate manuscripts [3, 9].

Copper-organic complexes can also result from the interaction of copper mineral pigments such as malachite (green) and azurite (blue) with the paint media [10]. Examples are copper complexes with fatty acids[3] or *copper soaps* [11–13] and copper oxalates [14, 15], whose precipitation causes efflorescence, darkening and delamination. Copper soaps have also been used as pigments [16].

Synthetic pigments made of copper-organic complexes are also susceptible to degradation due to interactions with the environment and paint media [17–19] hindering their identification.

In addition to painted surfaces, copper-organic complexes have been identified in archaeological composite objects (e.g. where metal fittings have been in contact with leather [20] and wood [21]), in sculptures coated with oils [22], waxes [23], and on painted copper surfaces [24].

Copper-organic complexes also play a role in the preservation of fibres found amidst the corrosion layers covering archaeological copper alloy objects that have been in contact with textiles. The preservation of organic material by copper corrosion has routinely been attributed to the biocidal properties of copper [25] and the mineralization of organic matter by inorganic copper salts [26, 27]. Laboratory experiments and data from archaeological samples demonstrate that in mineralized cellulose fibres such as cotton and linen, copper ions coordinate with hydroxyl groups. In protein fibres such as silk and wool, copper coordinates with carboxyl and amide groups [28, 29].

The spontaneous coordination of copper with organic substances outside archaeological/historical timeframes can be further illustrated with two examples from object conservation. The first is the recognition of the corrosive effect of organic acids that can emanate from storage and display materials, which manifests itself through the formation of copper acetates, copper formates and related compounds [3, 30]. The second example exploits the stability of copper-organic complexes as corrosion inhibitors. Whilst fatty acids, azoles, amines, amino acids and their derivatives have all been used as corrosion inhibitors, the most popular for archaeological copper is benzotriazole, which coordinates with Cu(I) and Cu(II) ions [31].

Except for copper oxalates, acetates, formates and some copper soaps [32], limited information is available in the cultural heritage and archaeological science literature about copper-organic compounds compared to inorganic ones. In the absence of commercial standards, studies of copper-organic complexes in cultural heritage often include the synthesis and characterization of reference compounds. However, the structural variability of copper-organic complexes increases for organic ligands that have more than one area of coordination (e.g. unsaturated bonds and multiple donor atoms). The work presented in the following chapters illustrates these constraints.

---

[3] Carboxylic acids with long carbon chains, e.g. palmitic acid with contains 16 carbons.

References

1. Greenwood NN, Earnshaw A (1997) 28—Copper, silver and gold. In: Greenwood NN, Earnshaw A (eds) Chemistry of the elements. Butterworth-Heinemann, Boston, pp 1173–1200
2. Greenwood NN, Earnshaw A (1997) 19—Coordination and organometallic compounds. In: Greenwood NN, Earnshaw A (eds) Chemistry of the elements. Butterworth-Heinemann, Boston, pp 905–943
3. Scott DA (2002) Chapter 9—The organic salts of copper. In: Scott DA (ed) Copper and bronze in art: corrosion., colorants, conservation. Getty Conservation Institute, Los Angeles, pp 268–316
4. Scott DA, Taniguchi Y, Koseto E (2001) The verisimilitude of verdigris: a review of the copper carboxylates. Stud Conserv 46:73–91
5. de la Roja JM, Baonza VG, San Andrés M (2007) Application of Raman microscopy to the characterization of different verdigris variants obtained using recipes from old treatises. Spectrochim Acta A Mol Biomol Spectrosc 68(4):1120–1125
6. Kühn H (1970) Verdigris and copper resinate. Stud Conserv 15(1):12–36
7. Colombini MP, Lanterna G, Mairani A et al (2001) Copper resinate: preparation, characterisation and study of degradation. Ann Chim 91(11–12):749–757
8. Conti C, Striova J, Aliatis I et al (2014) The detection of copper resinate pigment in works of art: contribution from Raman spectroscopy: detection of copper resinate pigment in works of art. J Raman Spectrosc 45(11–12):1186–1196
9. Scott DA, Khandekar N, Schilling MR et al (2001) Technical examination of a fifteenth-century German illuminated manuscript on paper: a case study in the identification of materials. Stud Conserv 46(2):93–108
10. Gunn M, Chottard G, Rivière E et al (2002) Chemical reactions between copper pigments and oleoresinous media. Stud Conserv 47(1):12–23
11. Poli T, Piccirillo A, Nervo M et al (2017) Interactions of natural resins and pigments in works of art. J Colloid Interface Sci 503(1):1–9
12. Noble P (2019) Chapter 1. A brief history of metal soaps in paintings from a conservation perspective. In: Casadio F, Keune K, Noble P et al (eds) Metal soaps in art. Springer, Heidelberg, pp 1–22
13. Boon JJ, Hoogland F, Keune K (2007) Chemical processes in aged oil paints affecting metal soap migration and aggregation. In: Parkin M (ed) AIC Paintings specialty group postprints, providence, Rhode Island, 16–19 June 2006. AIC, Rhode Island, pp 16–23
14. Bordignon F, Postorino P, Dore P et al (2008) The formation of metal oxalates in the painted layers of a medieval polychrome on stone, as revealed by micro-Raman spectroscopy. Stud Conserv 53(3):158–169
15. Nevin A, Melia J, Osticioli I et al (2008) The identification of copper oxalates in a 16th century Cypriot exterior wall painting using micro FTIR, micro Raman spectroscopy and gas chromatography-mass spectrometry. J Cult Herit 9(2):154–161
16. Moffatt EA, Sirois PJ, Miller J (1997) Analysis of the paints on a selection of Naskapi artifacts in ethnographic collections. Stud Conserv 42(2):65–73
17. Ioakimoglou E, Boyatzis S, Argitis P et al (1999) Thin-film study on the oxidation of linseed oil in the presence of selected copper pigments. Chem Mater 11(8):2013–2022
18. Altavilla C, Ciliberto E (2006) Copper resinate: an XPS study of degradation. Appl Phys A Mater Sci Process 83(4):699–703
19. Alter M, Binet L, Touati N et al (2019) Photochemical origin of the darkening of copper acetate and resinate pigments in historical paintings. Inorg Chem 58(19):13115–13128
20. Werner U, Selwyn LS, Stone T et al (2012) The removal of metal soaps from brass beads on a leather belt. Stud Conserv 57(1):20–23
21. Wang Q, Huang H, Shearman F (2009) Bronzes from the Sacred Animal Necropolis at Saqqara, Egypt: a study of the metals and corrosion. The British Museum—Tech Res Bull 3:73–82

22. Schrenk JL (1994) The royal art of Benin: surfaces, past and present. In: Scott DA, Podany J, Considine BB (eds) Ancient art and historic metals: conservation and scientific research: proceedings of a symposium organized by the J. Paul Getty Museum and the Getty Conservation Institute, November 1991. Getty Conservation Institute, Marina del Re, pp 51–62

23. Burmester A, Koller J (1987) Known and new corrosion products on bronzes: their identification and assessment, particularly in relation to organic protective coatings. In: Black J (ed) Recent advances in the conservation and analysis of artifacts: jubilee conservation conference papers. University of London. Institute of Archaeology, Summer Schools Press, London, pp 97–104

24. Pavlopoulou L, Watkinson DE (2006) The degradation of oil painted copper surfaces. Stud Conserv 51:55–65

25. Borkow G, Gabbay J (2005) Copper as a biocidal tool. Curr Med Chem 12(18):2163–2175

26. Janaway RC (1985) Dust to dust: the preservation of textile materials in metal artefact corrosion products with reference to inhumation graves. Sci Arch 27:29–34

27. Janaway RC (1989) Corrosion preserved textile evidence: mechanism, bias and interpretation. In: Janaway R, Scott B (eds) Evidence preserved in corrosion products: new fields in artefact studies. Proceedings of a joint conference between UKIC Archaeology Section and the Council for British Archaeology Science Committee, Leeds 1983. Institute for Conservation of Historic and Artistic Works, London, pp 21–29

28. Gillard RD, Hardman SM, Thomas RG et al (1994) The mineralization of fibres in burial environment. Stud Conserv 39(2):132–140

29. Gillard RD, Hardman SM (1996) Chapter 14—Investigation of fiber mineralization using Fourier transform infrared microscopy. In: Archaeological chemistry. ACS symposium series, vol 625. American Chemical Society, pp 173–186

30. Korenberg C (2006) Corrosion on metallic tokens stored in polyurethane foam. Stud Conserv 51(1):1–10

31. Scott DA (2002) Chapter 12. Conservation treatments for bronze objects. In: Scott DA copper and bronze in art: corrosion, colorants, conservation. Getty Conservation Institute, Los Angeles, pp 352–397

32. Robinet L, Corbeil M-C (2003) The characterization of metal soaps. Stud Conserv 48(1):23–24

Chapter 2
Corroding Copper in the Laboratory

An electrolytic cell is a medium where chemical reactions are produced through the application of electricity. It consists of two electrodes connected to a power supply and immersed in a conducting liquid (usually an aqueous medium) called an electrolyte.

The idea of using an electrolytic cell to produce copper corrosion containing organics and for the synthesis of copper-organic complexes was inspired by recipes to produce verdigris—the corrosion product of copper by acetic acid [1]—coupled with a desire to keep wet chemistry steps to a minimum. Electrolytic cells are fast and inexpensive, and the type of compounds produced is a factor of the composition of the electrolyte and the voltage applied to the system.

The typical electrolytic cell arrangement used in this research consisted of electrodes made of copper coupons[1] connected to a power supply[2] (Fig. 2.1).

The surface of the cathode coupon was abraded on both sides with a fibreglass pen to remove any pre-existing layer of copper oxide. Both coupons were washed with deionized water[3] and degreased with acetone prior to immersion in the electrolyte.

2.1 Proof-of-Concept Experiments

Before attempting to create copper-organic complexes, the electrolytic cell was initially set up to produce inorganic corrosion products. For this experiment (Experiment 1), an electrolyte composed of 0.02M Na_2CO_3 and 0.01M NaCl in deionized water was used. The power supply was adjusted to maintain a minimum anodic

---

[1] Alfa Aesar™ copper foil, 0.127 mm (0.005 in.) thick, annealed, 99.9% (metals basis) purchased from Fisher Scientific UK Ltd.

[2] Thurlby Thandar TSX1820 Precision DC Power supply.

[3] Obtained from a milli-Q reverse osmosis system (typically 18.2 Ω resistivity and <4 ppb carbon).

© The Author(s), under exclusive license to Springer Nature Switzerland AG 2022
L. C. Carvalho, *Beyond Copper Soaps*,
SpringerBriefs in Applied Sciences and Technology,
https://doi.org/10.1007/978-3-030-97892-1_2

Fig. 2.1 The electrolytic
cell setup: the power supply
is driving electrons from the
anode to the cathode causing
the (+) coupon to oxidize.
The positively-charged Cu
ions formed on the surface of
the anode attract
negatively-charged ions
present in the electrolyte

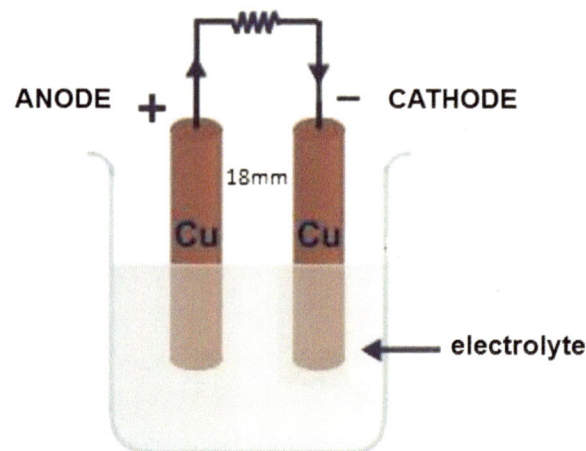

current of 0.01A passing through the system, to promote the slow release of copper
ions from the (+) coupon. The electrolyte was not agitated to allow for the deposition
of residues on the surface of the (+) coupon.

The pH of the electrolyte at the start of the experiment was 10.5. After a few
seconds from the power supply being switched on, both coupons started "fizzing",
with blue deposits forming along the edges of the (+) coupon, later spreading to other
areas. The blue deposits on the (+) coupon were eventually covered by a dark deposit
after 48 h, when the experiment was stopped. The coupons were carefully removed
from the electrolyte, washed with deionized water and allowed to dry in a fume hood
for seven days. The pH of the electrolyte was re-measured; it remained unchanged.

The mineral phases present in the corrosion deposit were identified by powder
X-ray diffraction (XRD) (Table 2.1).

Although some of the peaks have remained unassigned, the deposit was identified
as a mixture containing malachite $[Cu(CO_3) \cdot Cu(OH)_2]$, cuprite (Cu_2O) and tenorite
(CuO). The experiment was repeated three times, resulting in a residue containing
the same compounds in the same relative proportions.

Malachite and tenorite are typical corrosion products of copper in seawater, with
the formation of cuprite (as opposed to cupric chloride) due to the alkaline pH of the
electrolyte [2, 3]. Given these results, some of the possible reactions that could have
occurred during Experiment 1 are:

$$Anode\ (+)\quad 2Cu + 2OH^- \rightleftharpoons Cu_2O + H_2O + 2e^-$$
$$Cu + H_2O \rightleftharpoons CuO + H_2(gas)$$
$$2Cu^{2+} + 2OH^- + CO_3^{2-} \rightleftharpoons Cu(CO_3) \cdot Cu(OH)_2$$
$$Cu \rightleftharpoons Cu^{2+} + e^-$$
$$Cu \rightleftharpoons Cu^+ + e^-$$
$$2Cl^- \rightleftharpoons Cl_2(gas) + 2e$$

2θ (°)	d-spacings (nm)	I/I° (%)	Mineral phase
14.8483	5.9676	12.76	Malachite
16.2469	5.4558	11.55	
17.6243	5.0324	32.47	Malachite
24.0653	3.6981	42.59	Malachite
31.2190	2.8651	62.34	Malachite
32.2690	2.7742	42.74	Malachite
35.6540	2.5182	67.62	Tenorite
36.3704	2.4702	100.00	Cuprite
38.5509	2.3354	47.57	Tenorite
42.3216	2.1356	40.23	Cuprite
43.3031	2.0895	88.54	Copper
50.4232	1.8099	43.32	Copper
54.8286	1.6744	18.18	
58.0789	1.5882	19.50	
61.4149	1.5097	30.59	Cuprite
66.0172	1.4152	16.28	
67.8739	1.3809	12.53	
73.4421	1.2894	16.90	Cuprite
74.1092	1.2783	23.23	Copper

Table 2.1 XRD peak list and phase identification for Experiment 1 corrosion by comparing the sample's measured peak positions (2θ°) and their relative intensities (I/I°) to measured patterns of reference compounds

$$\text{Cathode } (-) \quad 2H_2O + 2e^- \rightleftharpoons H_2(\text{gas}) + 2OH^-$$
$$Na^+ + e^- \rightleftharpoons Na$$

For the next experiment (Experiment 2), n-hexadecanoic acid (aka palmitic acid) was added to the electrolyte. Palmitic acid was chosen because its reaction with copper could be traced by XRD and Fourier Transform Infrared (FTIR) and is, along with n-octadecanoic acid (aka stearic acid), ubiquitous in organic residues recovered from archaeological ceramics.

Given palmitic acid's low solubility in water, the compound was pre-dissolved in warm methanol and the solution incorporated into an aqueous electrolyte containing sodium, carbonate and chloride ions. The final composition of the electrolyte used in Experiment 2 was 0.02M Palmitic acid, 0.01M Na_2CO_3 and 0.01M NaCl in 50:50 v/v methanol/deionized water and its pH adjusted to 10.5 with a few drops of 1M NaOH.

Upon combining the palmitic acid methanolic solution to the acqueous electrolyte, a white flocculent substance formed, rising to the surface of the electrolyte. This substance was probably sodium palmitate, the first of a two-step method for the synthesis of divalent metal soaps [4, 5]. After 48 h, the coupons were carefully removed, washed with ethanol to remove excess acid and allowed to dry for seven days in a fume cupboard.

Table 2.2 XRD peak list and phase identifications for Experiment 2 corrosion

2θ (°)	d-spacings (nm)	I/I° (%)	Mineral phase
6.1505	1.4370	9.52	Copper palmitate
8.2511	1.0716	0.66	Copper palmitate
10.2987	0.8589	3.93	Copper palmitate
15.4371	0.5740	17.68	
16.1899	0.5475	29.96	Atacamite
21.1714	0.4167	23.07	
22.4314	0.3963	24.72	
29.4484	0.3033	26.90	
32.1716	0.2782	39.42	
36.3971	0.2468	100.00	Cuprite
39.6965	0.2271	31.77	Atacamite
42.2995	0.2137	40.18	Cuprite
43.3024	0.2090	36.70	Copper
50.5016	0.1807	19.94	Copper
61.3207	0.1512	25.29	Cuprite
67.7420	0.1383	7.89	
73.5524	0.2877	11.14	Cuprite
74.12.09	0.1278	10.20	Copper

The dried blue deposit covering the (+) copouns was removed, homogenized and analysed by powder XRD alongside a copper palmitate reference standard (synthetized according to the protocol described in the Appendix). The mineral phases identified in the Experiment 2 corrosion were atacamite and cuprite, with a few peaks at low 2θ angles, assigned to copper palmitate (Table 2.2).

The presence of copper palmitate amongst the Experiment 2 corrosion was confirmed by FTIR (Fig. 2.2).

Of interest are the strong peak at 1580 cm^{-1} being the COO$^-$ asymmetric stretch characteristic of all metal soaps [6, p. 28] and additional peaks from atacamite [7] in the 3,500–3,100 cm^{-1} and 1,000–800 cm^{-1} regions of the FTIR spectrum, formed due to the presence of chloride ions in the electrolyte. The possible reactions that occurred during Experiment 2 are:

$$\text{Anode (+)} \quad 2\text{Cu} + \text{H}_2\text{O} \rightleftharpoons \text{Cu}_2\text{O} + 2\text{H}^+ + 2\text{e}^-$$
$$2\text{Cu} + 3\text{H}_2\text{O} + \text{Cl}^- \rightleftharpoons \text{Cu}_2(\text{OH})_3\text{Cl} + 3\text{H}^+ + 2\text{e}^-$$
$$\text{Cu}^{n+} + n\text{RCOONa} \rightleftharpoons (\text{RCOO})n\text{Cu} + n\text{Na}^+$$
$$\text{Cu}^{n+} + n\text{RCOO}- \rightleftharpoons (\text{RCOO})n\text{Cu}$$
$$\text{Cu} \rightleftharpoons \text{Cu}^{2+} + \text{e}^-$$
$$\text{Cu} \rightleftharpoons \text{Cu}^+ + \text{e}^-$$

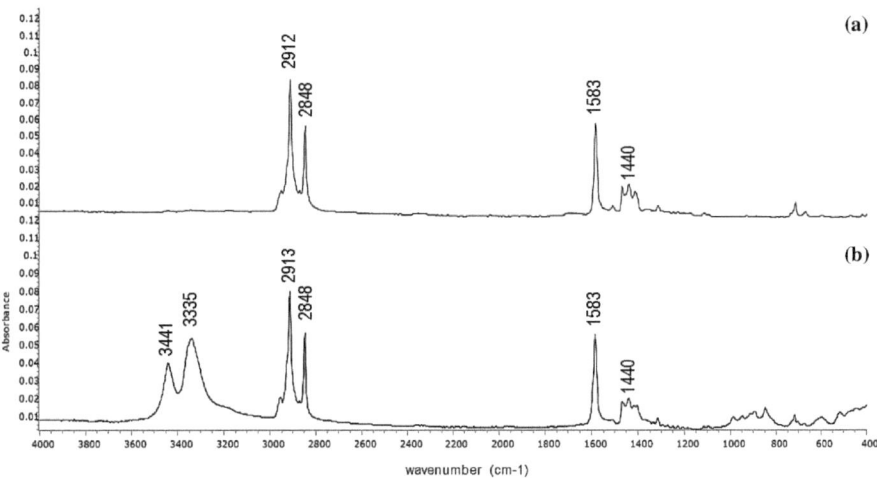

Fig. 2.2 FTIR spectra of Experiment 2 corrosion (**a**) and copper palmitate (**b**)

$$\text{Cathode } (-) \quad 2e^- + 2H_2O \;\rightarrow\; H_2(\text{gas}) + 2OH^-$$
$$e^- + Na^+ \rightarrow Na$$
$$Cu^{2+} + 2e^- \rightarrow Cu$$

To test the effect of the pH in the composition of the corrosion product, Experiment 2 was repeated at neutral pH. Analysis of the resulting corrosion product formed on the (+) coupon by FTIR yielded a spectrum (not shown) with a carboxylate ion peak at 1580 cm^{-1}, indicating that copper can react with palmitic acid in both neutral and alkaline environments.

2.2 Synthesis of Copper-Organic Complexes

To test if the electrolytic cell could be used to generate copper complexes with other types of organic ligands, the experimental conditions were modified as follows:

I.   Composition of the electrolyte: 0.01M organic compound and 0.01M NaCl or ammonium acetate in 50:50v/v co-solvent (methanol or acetone)/deionized water,
II. anodic current set to 0.02A to speed up copper ionization; and
III. electrolyte agitated to stimulate reaction with the organic ligand.

Different organic substances were used in these electrochemical synthesis experiments (Table 2.3) selected for containing potential metal chelating sites.

Table 2.3 Organic compounds used in the copper complexation experiments

Type	Compound	Co-solvent	Electrolyte
Aromatic	Phenol	Methanol	Ammonium acetate
Carbohydrate	Lactose	Methanol	Ammonium acetate
Carboxylic acid (saturated)	Palmitic and Stearic acids	Methanol	Sodium chloride
Carboxylic acid (unsaturated)	Oleic and Linoleic acids	Acetone	Ammonium acetate
Diacylglycerol	Glyceryl palmitate	Methanol	Ammonium acetate
Ester	Ethyl palmitate	Methanol	Ammonium acetate
Phenolic acid	Gallic acid	Methanol	Ammonium acetate
Phosphoprotein	Bovine casein	Methanol	Ammonium acetate
Polyol	Glycerol	Methanol	Ammonium acetate
Polyphenol	Resveratrol	Methanol	Ammonium acetate
Triacylglycerol	1,3-dipalmitol 2-oleyglycerol	Methanol	Ammonium acetate

In all experiments, the organic compound was pre-dissolved in a co-solvent and heated in a bain-marie to aid dissolution as required. The pH of the electrolyte was adjusted to 10.5 with 1M NaOH and the experiments were run for 1 h.

At the end of the experiment, the coupons were removed and washed with the co-solvent. The electrolyte was vacuum filtered on fibreglass filter paper and the precipitate washed with the co-solvent. The copper coupons and paper filter containing the residue were left to dry in a fume cupboard for seven days.

The experiments with electrolytes containing phenol, glyceryl 1,3-dipalmitate, ethyl palmitate, gallic acid, glycerol, resveratrol and 1,3-dipalmitol-2-oleyglycerol did not yield a deposit over the (+) coupon nor a precipitate that could be recovered from the paper filter.

Deposits over the (+) coupon were obtained in the experiments with carboxylic acids, lactose and casein. Their characterization is presented in Chap. 3.

References

1. Scott DA, Taniguchi Y, Koseto E (2001) The verisimilitude of verdigris: a review of the copper carboxylates. Stud Conserv 46:73–91

2. Bianchi G, Longhi P (1973) Copper in seawater, potential-pH diagrams. Corros Sci 13(11):853–864
3. Alfantazi AM, Ahmed TM, Tromans D (2009) Corrosion behavior of copper alloys in chloride media. Mater Eng 30(7):2425–2430
4. Koenig AE (1914) On the stearates and palmitates of the heavy metals with remarks concerning instantaneous precipitations in insulating solutions. JACS 36:951–961
5. Corbeil M-C, Robinet L (2002) X-ray powder diffraction data for selected metal soaps. Powder Diffr 17(1):52–56
6. Robinet L, Corbeil M-C (2003) The characterization of metal soaps. Stud Conserv 48(1):23–40
7. https://rruff.info/Atacamite. Accessed 17 Dec 2021

Chapter 3
Copper-Organic Complexes Synthetized Electrochemically

The anode coupons at the end of the experiments with carboxylate acids, lactose and casein contained deposits varying in shades of blue/green (Fig. 3.1).

3.1 Characterization of Copper Soaps

The (+) coupons from the electrolyte containing palmitic and stearic acids were covered by a light greenish/blue powdery deposit (Fig. 3.1a, b) whilst those exposed to oleic and linoleic acids were covered by a dark green, reflective sticky film (Fig. 3.1c, d). For ease of reference, these deposits will be referred to as Cu-Palmitate, Cu-Stearate, Cu-Oleate and Cu-Linoleate, respectively.

3.1.1 Fourier Transform Infrared Spectroscopy (FTIR)

Carboxylic acids normally exist as dimers due to strong intermolecular hydrogen bonding [1] characterized by hydroxyl (OH) and carbonyl (C=O) vibrations. Their most recognizable OH vibrations are a broad band in the $3300–2500 \text{ cm}^{-1}$ region and a variable intensity band around $955–890 \text{ cm}^{-1}$, whilst the C=O vibration appears as a strong band within the $1715–1680 \text{ cm}^{-1}$ region. When ionization occurs forming the COO^- carboxylate ion, resonance is possible between the two C–O bonds [2]. The C=O band is replaced by two vibrations: a characteristic strong band between 1610 and 1550 cm^{-1} (COO^- asymmetric stretching) and a medium band between 1420 and 1300 cm^{-1} (COO^- symmetric stretching) [3].

© The Author(s), under exclusive license to Springer Nature Switzerland AG 2022   13
L. C. Carvalho, *Beyond Copper Soaps*,
SpringerBriefs in Applied Sciences and Technology,
https://doi.org/10.1007/978-3-030-97892-1_3

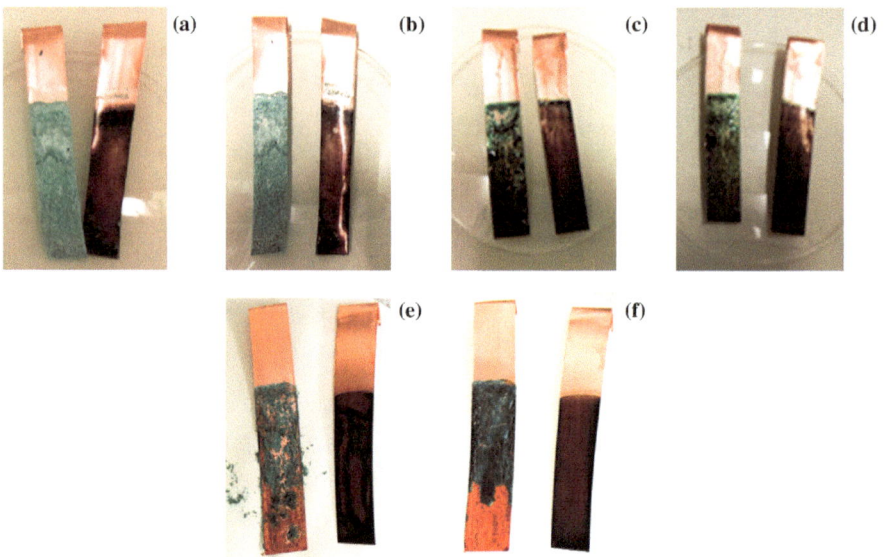

Fig. 3.1 Copper coupons at the end of experiments (anode coupons are those on the left) following exposure to palmitic acid (**a**), stearic acid (**b**), oleic acid (**c**), linoleic acid (**d**), lactose (**e**) and casein (**f**)

The COO$^-$ bands in Cu-Palmitate appear as a strong band around 1583 cm^{-1} and a medium band around 1422 cm^{-1}. The appearance of these bands coincides with the reduction of the C=O band around 1700 cm^{-1} (Fig. 3.2b).

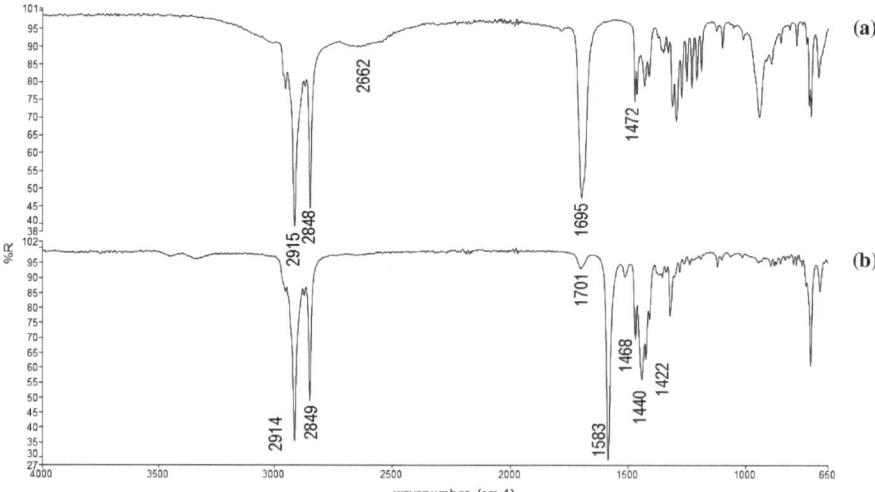

Fig. 3.2 FTIR spectra of Palmitic Acid (**a**) and Cu-Palmitate (**b**) obtained in ATR mode

Table 3.1 Characteristic FTIR bands of copper soaps

Cu-Palmitate	Cu-Stearate	Cu-Oleate	Cu-Linoleate	Band assignment
2914 vs	2912 vs	2916 vs	2924 vs	C–H stretching vibrations
2849 vs	2846 vs	2849 s	2853 s	C–H stretching vibrations
1701 w	1695 w, br		1729 m	C=O (dimer) stretching
1583 vs	1585 vs	1583 vs	1601 vs	COO⁻ asymmetric stretching vibrations
1422 m	1419 m	1414 m	1413 m	COO⁻ symmetric stretching vibrations
946 vw				OH deformation
721 m	716 m	718 w		COO⁻ scissor vibrations
682 m	673 m			COO⁻ wagging vibrations

vs (very strong), s (strong), m (medium), w (weak)

Analogous bands to Cu-Palmitate were also identified in the spectra of the remaining copper soaps (Table 3.1).

The peaks at around 1700 cm^{-1} and at 950 cm^{-1} are due to the presence of acid residues—unreacted acid or resulting from the behaviour of the carboxylate ions in an aqueous medium [4] according to the following reactions:

$$2H_2O \rightleftharpoons H_3O^+ + OH^-$$

$$RCOO^- + H_3O^+ \rightleftharpoons RCOOH + H_2O$$

The (+) coupons containing the copper soaps were also analysed with an ATR microscope attachment (Fig. 3.3). The main advantage of using an ATR microscope is that it enables spatially-resolved analysis without sample removal, ideal for the

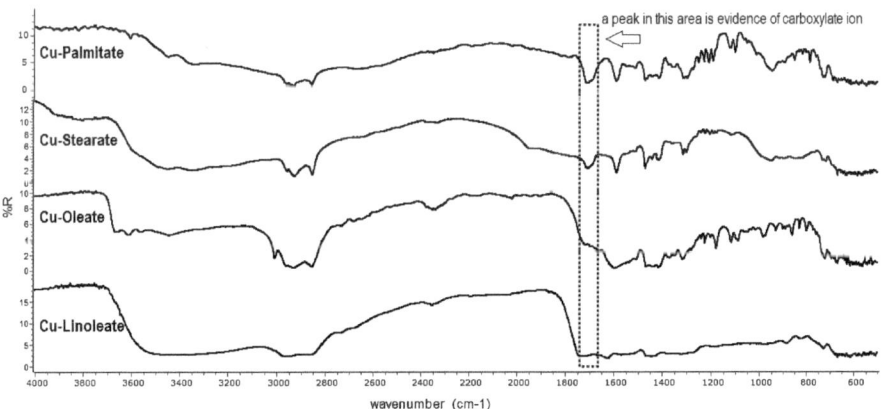

Fig. 3.3 FTIR spectra of copper soaps obtained with ATR microscope

study of paintings and manuscripts. The obtained spectra for Cu-Palmitate and Cu-Stearate contain a well-defined peak corresponding to asymmetric COO⁻ stretching vibrations (area within a rectangle in Fig. 3.3), practically non-existent in Cu-Oleate and Cu-Linoleate spectra. This may explain why unsaturated copper soaps are not reported in FTIR-ATR analyses in the literature [5–7].

3.1.2 Raman Spectroscopy

Theoretical and experimental studies have assigned bands below 590 cm^{-1} in the infrared and Raman spectra of metal–organic complexes to metal-nitrogen (M–N) and metal–oxygen (M–O) vibrations [1, 3]. Cu–O vibrations have been specifically assigned to bands in the 350-180 cm^{-1} region for copper(II) acetate [8, 9], copper(II) hexanoate [10] and copper palmitate and copper stearate [11]. The bands at 307 cm^{-1} and 255 cm^{-1} in Cu-Palmitate's spectrum (Fig. 3.4b), which do not appear in the spectrum of palmitic acid (Fig. 3.4a), may be attributed to Cu–O vibrations and, thus, considered evidence of metal–organic complexation. Similar bands were also identified in the spectra of Cu-Stearate and Cu-Oleate but not in Cu-Linoleate due to fluorescence (Fig. 3.5d).

The main difference between the copper soaps' spectra is the broad peak at around 1630 cm^{-1} in Cu-Oleate and Cu-Linoleate (Fig. 3.5c, d and Table 3.2). This peak corresponds to the C=C bond(s), which are very weak in FTIR.

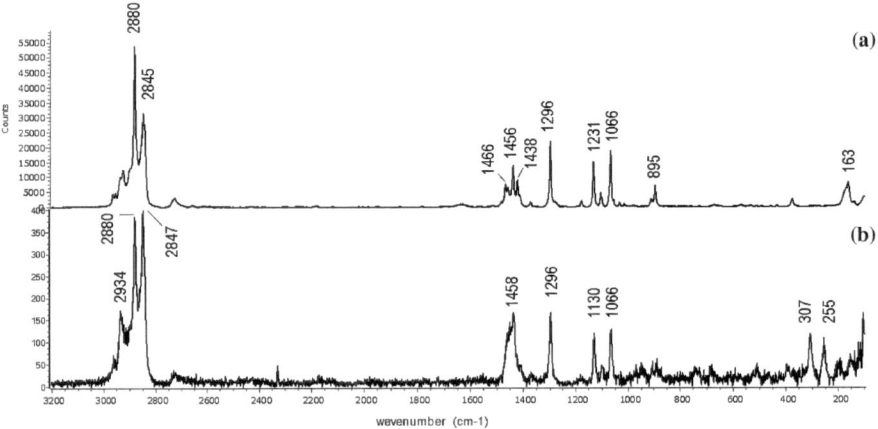

Fig. 3.4 Raman spectra of palmitic acid (**a**) and Cu-Palmitate (**b**) baseline corrected

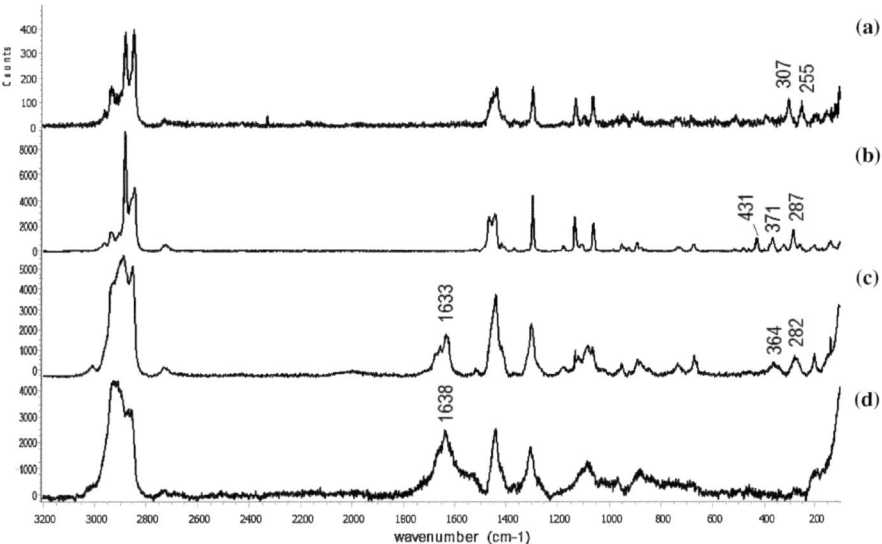

Fig. 3.5 Raman spectra of Cu-Palmitate (**a**), Cu-Stearate (**b**), Cu-Oleate (**c**) and Cu-Linoleate (**d**)

Table 3.2 Characteristic Raman bands of copper soaps

Cu-Palmitate	Cu-Stearate	Cu-Oleate	Cu-Linoleate	Band assignments
2934 m	2931 m	3007 m	2920 s	C–H stretching vibrations
2880 vs	2880 vs	2867 s	2867 s	C–H stretching vibrations
2847 vs	2847 vs			C–H stretching vibrations
		1633 s	1638 s	C=C stretching vibrations
1556 vw	1523 vw	1517 w		COO$^-$ asymmetric stretching vibration
1437 s	1442 s	1440 s	1440 s	CH$_2$ scissors aliphatic
1296 s	1297 s	1301 s	1302 s	CH$_2$ torsion
1130 s	1132 s	1132 s		C–C vibrations
1066 s	1063 s	1070 s	1085 s	COO$^-$ deformation
	431 m			Cu–O vibration
	371 m	364 m		Cu–O vibration
307 m				Cu–O vibration
	287 m	282 m		Cu–O vibration
255 m				Cu–O vibration

vs (very strong), s (strong), m (medium), w (weak), vw (very weak)

3.1.3   X-ray Photoelectron Spectroscopy (XPS)

The elemental composition obtained by XPS shows an agreement between theoretical and experimental values for Cu-Palmitate and Cu-Oleate, but deviations for Cu-Stearate of 11% and of 61% for Cu-Linoleate (Table 3.3). The high deviation obtained for Cu-Linoleate is likely to result from the decomposition of the sample prior to or during analysis, but invisible to the naked eye.

Cu(II) ions can be differentiated from Cu(I) ions, independent of the ligand, by the presence of shake-up (or satellite) peaks in their Cu2p spectrum in the regions below 960 eV and between 945 and 940 eV [12–14]. The Cu2p spectra for all copper soaps analysed contain satellite peaks in these regions confirming the presence of Cu(II) ions.

Evidence of metal complexation can also be obtained by comparing the high-resolution spectra of C1s and O1s regions of the carboxylic acid and its copper soap.

Palmitic acid's C1s spectrum contains two well-defined peaks (Fig. 3.6a). One peak is centred at 284.7 eV corresponding to C–C/C–H bonds in a carbon chain and another at 289.1 eV corresponding to a carbon coordinated to oxygen atoms, O–C=O. In contrast, the C1s spectrum for Cu-Palmitate (Fig. 3.6b) shows a single large peak. This peak can be decomposed into a peak at 284.8 eV corresponding to C–C/C–H and two further peaks for the carboxylate carbon: one peak at 288.2 eV can be assigned to C=O [15] and another peak at 286.2 eV to C–O [16].

The O1s spectra of palmitic acid (Fig. 3.6c) also differs from that obtained for Cu-Palmitate (Fig. 3.6d). The binding energy (BE) of the O=C decreases from 532.1 to 531.8 eV with an increase in the O–C from 533.2 to 533.3 eV. Similar values were also observed for the remaining copper soaps (Table 3.4).

As the position of an XPS peak for an atom is affected by the initial binding energy of the electron in its orbital, any changes in the atom's vicinity are likely to shift peak positions. However, copper soaps exist as coordination complexes where the bond between the metal ion and the organic component is often delocalized, without having a distinct covalent or ionic character. Hence, the chemical shifts imposed by the metal ion on the carboxylic function cannot be easily measured by this technique.

Table 3.3 Elemental Composition and Atomic Proportions for palmitic acid and copper soaps

	Ligand formula	Relative atomic %			C:O:Cu	
		C1s	O1s	Cu2p	Theoretical	Experimental*
Palmitic Acid	$C_{16}H_{32}O_2$	89.32	10.62	–	32:4:0	34:4:0
Cu-Palmitate	$C_{16}H_{32}O_2$	85.53	10.72	2.81	32:4:1	32:4:1
Cu-Stearate	$C_{18}H_{36}O_2$	84.62	10.62	2.93	36:4:1	32:4:1
Cu-Oleate	$C_{18}H_{34}O_2$	87.97	9.81	2.25	36:4:1	36:4:1
Cu-Linoleate	$C_{18}H_{32}O_2$	68.93	24.23	3.00	36:4:1	11:4:1

* Rounded to the closest integer

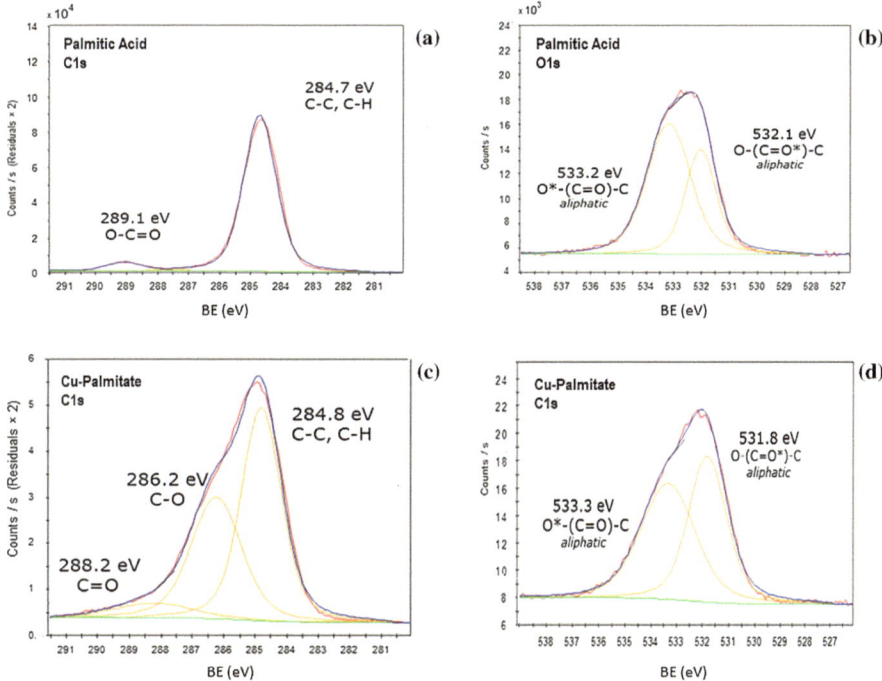

Fig. 3.6 Spectra of C1s and O1s regions of palmitic acid (**a**, **b**) and Cu-Palmitate (**c**, **d**)

Table 3.4 C1s and O1s peaks binding energies (BE) for copper soaps

Region	Chemical bonds	Cu-Palmitate	Cu-Stearate	Cu-Oleate	Cu-Linoleate
C1s BE (eV)	C–C, C–H, C=C	284.8	284.6	284.6	284.4
	C–O	286.2	285.8	285.9	285.9
	C=O, C–O–Cu	288.2	287.8	288.7	288.2
O1s BE (eV)	O–(C=O*)–C	531.8	531.7	531.7	531.6
	O*–(C=O)–C	533.3	533.1	533.0	532.8

3.1.4 X-ray Diffraction (XRD)

Diffraction patterns were obtained between 5 and 40° 2θ for Cu-Palmitate (Table 3.5), Cu-Stearate (Table 3.6) and Cu-Oleate (Table 3.7). No diffraction pattern was obtained for Cu-Linoleate.

Below 25° 2θ, the intensity and position of the peaks appear to correlate with the length of the carbon chain and its level of unsaturation. An increase in the carbon chain (from C16 in Cu-Palmitate to C18 in Cu-Stearate) resulted in a shift in peak positions to lower angles, analogous to the acid forms as reported in the literature [11].

Table 3.5 X-ray powder diffraction pattern obtained for Cu-Palmitate, most intense peaks in BOLD

2θ (°)	d-spacings (nm)	I/I° (%)	2θ (°)	d-spacings (nm)	I/I° (%)
5.991	**1.4739**	**28.3**	18.522	0.4787	1.3
6.246	**1.4139**	**100.0**	18.989	0.4670	1.5
7.435	**1.1880**	**34.7**	19.385	0.4575	3.5
8.353	1.0576	10.8	19.893	0.4460	3.4
8.812	1.0027	0.8	20.443	0.4341	1.9
9.196	0.9607	1.7	20.813	0.4265	2.1
9.901	0.8927	10.4	**21.612**	**0.4109**	**43.1**
10.462	0.8449	18.6	22.463	0.3955	1.5
11.129	0.7944	4.2	22.866	0.3886	3.7
11.739	0.7533	5.2	23.135	0.3842	10.8
11.878	0.7445	5.7	23.803	0.3735	4.4
12.402	0.7131	12.0	24.207	0.3674	22.4
13.315	0.6644	1.9	27.212	0.3276	1.0
13.755	0.6433	1.9	27.625	0.3227	0.6
14.548	0.6084	2.5	29.083	0.3068	1.0
15.799	0.6047	1.1	30.26	0.2951	3.0
17.367	0.5102	1.1			

Table 3.6 X-ray powder diffraction pattern obtained for Cu-Stearate, most intense peaks in BOLD

2θ (°)	d-spacings (nm)	I/I° (%)	2θ (°)	d-spacings (nm)	I/I° (%)
5.272	**1.6749**	**39.8**	13.145	0.6730	6.5
5.611	**1.5739**	**100.0**	14.071	0.6289	0.6
7.012	1.2596	9.0	15.047	0.5883	0.5
7.499	1.1779	14.5	15.846	0.5588	0.7
8.772	1.0073	10.2	16.188	0.5471	0.6
8.959	0.9862	3.9	16.94	0.5230	1.1
9.375	**0.9426**	**26.9**	19.385	0.4575	1.0
10.528	0.8396	2.2	19.752	0.4491	0.5
10.851	0.8147	0.8	20.196	0.4394	0.3
11.108	0.7959	1.6	21.378	0.4153	0.4
11.261	0.7852	3.6	23.01	0.3862	1.4
12.299	0.7191	3.3	23.565	0.3772	1.2
12.56	0.7042	1.4			

Table 3.7 X-ray powder diffraction pattern obtained for Cu-Oleate, most intense peaks in BOLD

2θ (°)	d-spacings (nm)	I/Io (%)	2θ (°)	d-spacings (nm)	I/I° (%)
6.194	**1.4257**	**100.0**	**19.823**	**0.4475**	**19.1**
8.265	1.0689	10.1	20.382	0.4354	3.1
10.388	**0.8509**	**37.0**	21.741	0.4085	3.8
10.87	0.8133	14.5	22.404	0.3965	3.9
11.532	0.7666	9.3	22.816	0.3895	3.1
12.438	0.7111	7.6	23.446	0.3791	5.1
14.525	0.6093	6.9	24.091	0.3691	3.0
18.752	0.4728	1.6	25.953	0.3430	0.8

The effect of carbon chain unsaturation on the diffraction patterns can be observed by comparing the patterns obtained for Cu-Stearate (Table 3.6) with those obtained for Cu-Oleate (Table 3.7). The highest intensity peak of Cu-Oleate has a d-spacing (1.4257 nm) in the same order of magnitude as that of Cu-Palmitate (1.4139 nm), which has a shorter carbon chain, as reported in the literature [2, 17].

The diffraction patterns obtained for the copper soaps conform to the general structural arrangement of a metal soap where metal atoms form a mono cationic layer sandwiched by the carboxylate carbon chains [18]. The effect of this arrangement is high-intensity long d-spacings at low angles, as the x-ray strikes the length of the metal soap dimer molecule, with the short d-spacings at higher angles corresponding to the chain packing arrangement [18].

3.1.5 Mass Spectrometry

3.1.5.1 Direct Mass Measurement by Flow Injection Analysis (FIA)

Only the mass of the carboxylate ions could be found, but at extremely low counts and varying according to the solvent (Table 3.8). The low abundance of carboxylate

Table 3.8 Copper soaps' relative solubility in selected solvents

	Ion m/z	Ion abundance in		
		Acetonitrile	Chloroform	Methanol
Cu-Palmitate	255.2327	*	***	**
Cu-Stearate	283.2636	**	**	***
Cu-Oleate	281.2119	**	***	*
Cu-Linoleate	279.2317	*	*	**

KEY: * trace ** low *** medium in logarithmic scale

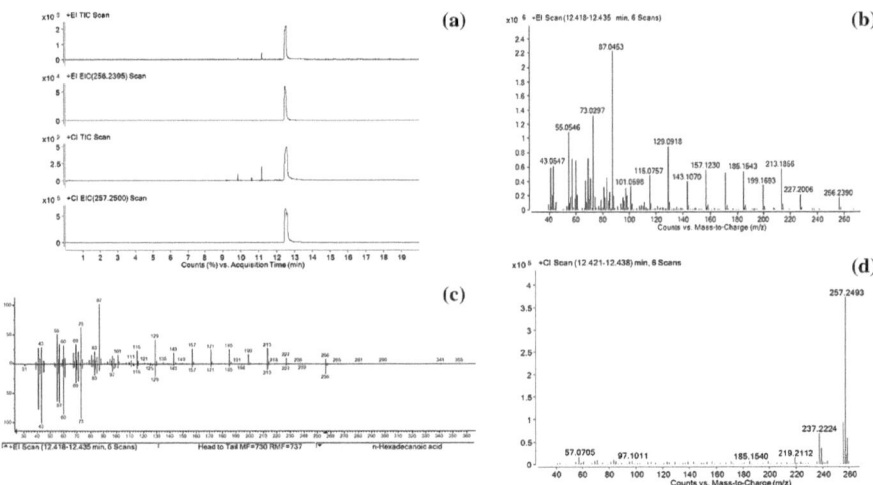

Fig. 3.7 Cu-Palmitate's EI and CI chromatograms (**a**), mass spectra in EI (**b**) and CI (**d**) and NIST library match (**c**)

ions may be a consequence of the strength of the bond between copper and carboxylate ligand. The low ion count for the carboxylate ions obtained in acetonitrile was expected as this solvent can chelate Cu(II) ions resulting in copper(I)-acetonitrile complexes [19]. Interestingly, copper(I)-acetonitrile complexes were identified in the Cu-Stearate (main ion peak) and Cu-Oleate. Both samples also yielded the highest relative amount of carboxylate ions.

3.1.5.2 Gas Chromatography with Quadrupole Time-of-Flight Mass Spectrometry using a Thermal Separation Probe (GC-QTOF-MS with TSP)

Despite the presence of minor impurities, the chromatograms of Cu-Palmitate (Fig. 3.7a), Cu-Stearate (Fig. 3.8a) and Cu-Oleate (Fig. 3.9a) all contained a large chromatographic peak corresponding to the carboxylic acid, both in Electron Ionization (EI) and Chemical Ionization (CI) modes. For Cu-Linoleate, the largest peak in its chromatograms (Fig. 3.10a) corresponds to azelaic acid ($C_9H_{16}O_4$) due to the oxidative cleavage of linoleic acid [20–22].

3.1.5.3 Solvent Extract Derivatized with N, O-Bistrifluoroacetamide (BSTFA) Analysed by Gas Chromatography with Mass Spectrometry (GC–MS)

Only the palmitic and stearic ligands (as methyl esters) were identified.

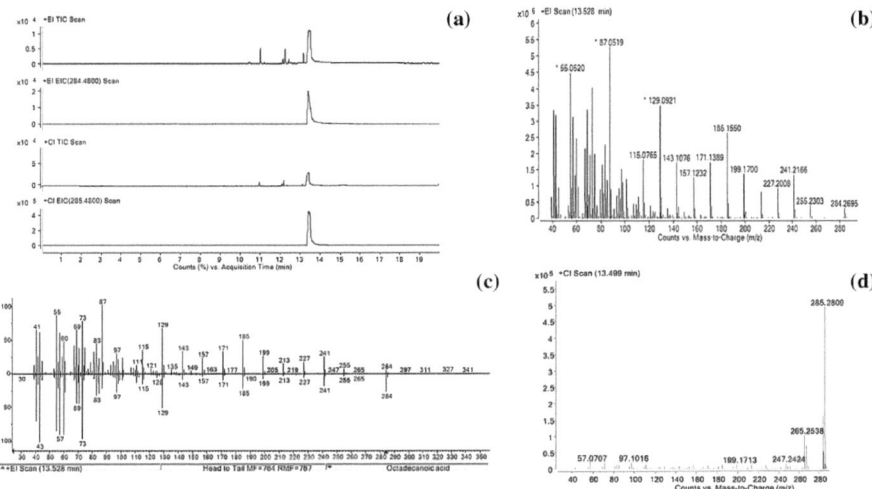

Fig. 3.8 Cu-Stearate's EI and CI chromatograms (**a**), mass spectra in EI (**b**) and CI (**d**) and NIST library match (**c**)

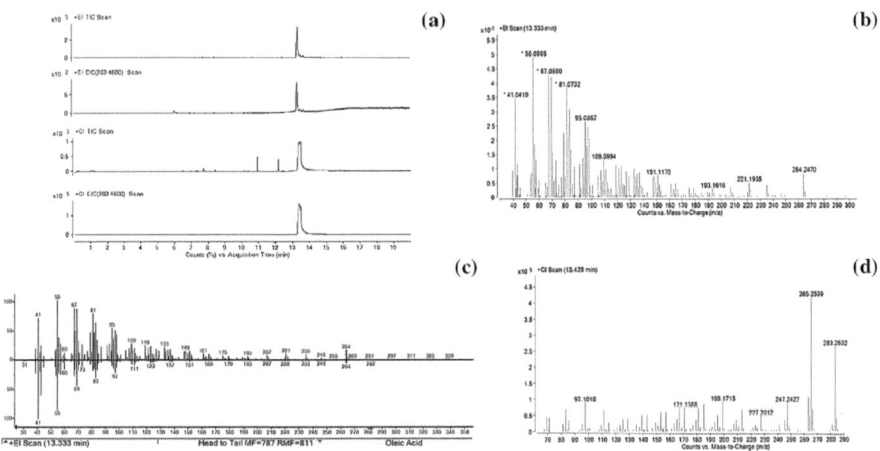

Fig. 3.9 Cu-Oleate's EI and CI chromatograms (**a**), mass spectra in EI (**b**) and CI (**d**) and NIST library match (**c**)

3.1.6 Discussion

The FTIR, Raman and XRD results largely agree with those published in the literature [11, 17]. Confirmation that copper is coordinating with oxygen ions was obtained by the strong peak at around 1580 cm^{-1} in the FTIR spectra and the peaks under 450 cm^{-1} in the Raman spectra.

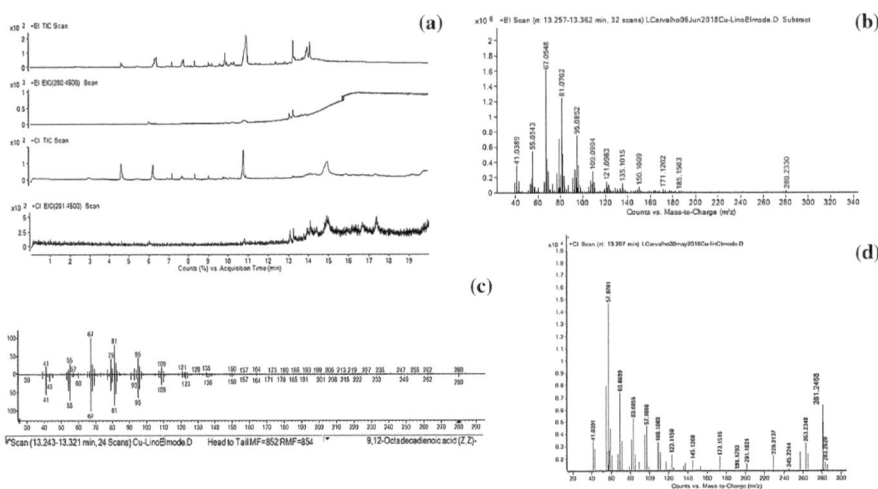

Fig. 3.10 Cu-Linoleate's EI and CI chromatograms (**a**), mass spectra in EI (**b**) and CI (**d**) and NIST library match (**c**)

The blue/green colour of the compounds indicated the presence of Cu(II) ions, which was confirmed by XPS. The XPS survey data also indicates that each Cu(II) ion coordinates with two carboxylate ions in a dimer arrangement, leading to the long d-spacings observed in the XRD data. This stoichiometry could not be confirmed for Cu-Linoleate, possibly because the sample was damaged/degraded before or during data acquisition [23].

The results obtained from the mass spectrometry analyses differed. Only the saturated ligands were identified using the solvent extraction protocol. The low solubility of metal soaps in organic solvents has long been recognized as a drawback to GC–MS protocols that include solvent extraction [24]. However, the results obtained indicate that the bond between copper ions and palmitate/stearate may have a more ionic character (making these soaps more soluble) than the bond with oleate/linoleate ligands.

3.2 Characterization of Cu-Lactose

The (+) coupon at the end of the experiment with the electrolyte containing lactose was covered by a blue powdery deposit (Fig. 3.1e), from this point onwards referred to as Cu-Lactose.

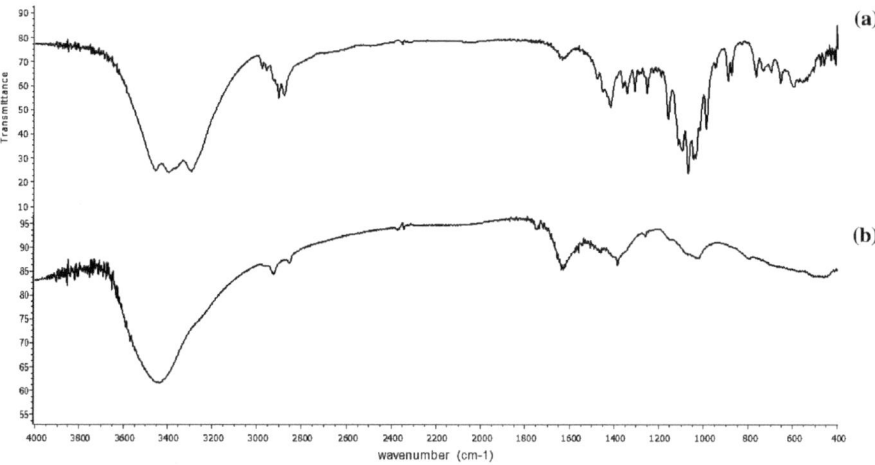

Fig. 3.11 FTIR spectra of Lactose (**a**) and Cu-Lactose (**b**) obtained by the KBr method

3.2.1   FTIR

Saccharides usually bind to a metal ligand via deprotonation of one or more hydroxyl groups [25, 26]. Lactose is a disaccharide consisting of one galactose unit linked to a glucose unit by an *O*-glycosidic bond. Given its many hydroxyl groups, many modes of coordination with metal ions are possible.

The spectra of lactose and Cu-Lactose are very similar (Fig. 3.11) with the most characteristic peak being the hydroxylated broad band centred around 3400 cm^{-1} (Table 3.9). For lactose, this band contains small peaks at 3453, 3392 and 3294 cm^{-1} due to its high level of crystallinity [27, 28]. In Cu-Lactose (Fig. 3.11b), these small peaks disappear due to the presence of water molecules [29] trapped within the complex during its synthesis. The three sharp bands at 1480–1300 cm^{-1}, 1160–890 cm^{-1} and 660–460 cm^{-1} shift to around 1400 cm^{-1}, 1080 cm^{-1} and 500 cm^{-1} due to interaction with copper ions [26, 29].

3.2.2   Raman

The shape of the spectrum of Cu-Lactose indicates fluorescence from the complex itself or impurities within it (Fig. 3.12b).

Fluorescence results from the emission of energy absorbed by the molecule as they return to a lower state of energy. If the laser excitation energy is close to the electronic transition energy of the molecule, fluorescence may obscure Raman scattering [30].

The most common method for overcoming fluorescence is to use a laser with lower wavelengths [30, 31], but none were available. Attempts were made to reduce the

Table 3.9 Main FTIR peaks for Lactose and Cu-Lactose with band assignments

Lactose		Cu-Lactose		Band Assignments
3744–3000	br	3694–3003	br	OH stretching vibration
2978	w			CH stretching vibrations
2957	w			CH stretching vibrations
2903	m			CH symmetric vibration of CH_2OH group
2878	m	2853	w	CH asymmetric vibration of CH_2OH group
1454	w			H-CH group scissoring vibrations
1417	m			CH_2 wagging (galactose)
1361	m			OCH deformation
1343	m			CCH deformation
1253	m			C–H rocking, OH rocking (glc)
1160	s	1160–890	br	C–O–C stretching vibration (glc)
1097	m			C–O–C stretching vibrations (gal)
1070	s			C–C stretching vibrations (gal)
990	m			C–H out-of-phase ring stretching and twisting
892	m			C–H out-of-phase ring stretching and twisting
597	m			OCO in-plane bending (gal)
461	m			C–C–O in-plane bending (glc)

br (broad), vs (very strong), s (strong), m (medium), w (weak), gal (galactose) and glc (glucose)

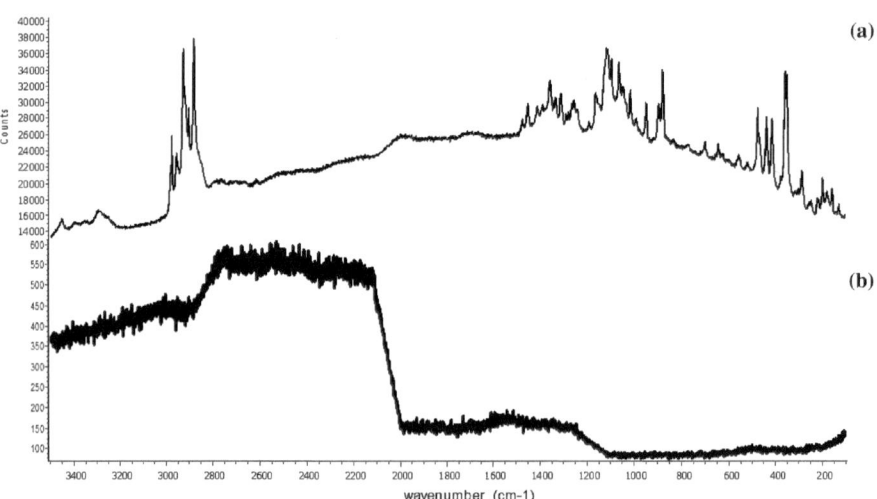

Fig. 3.12 Raman spectra of Lactose (**a**) and Cu-Lactose (**b**)

fluorescence by photobleaching the sample and by analysing Cu-Lactose directly off the copper coupon (as opposed to removing it to a glass slide)—both unsuccessful.

3.2.3 XPS

The elemental composition measured for lactose as relative atomic percentages was 52.4% carbon, 43.5% oxygen and 4.1% silicon. Despite the silicon contamination, these values are close to those reported for anhydrous α/β lactose [32]. For Cu-Lactose, the relative atomic percentages were 43.3% carbon, 41.0% oxygen and 15.7% copper; its Cu2p spectrum contained the characteristic satellite peaks indicating the presence of Cu(II) ions.

The C1s and O1s spectra for lactose and Cu-Lactose also differed (Fig. 3.13). In the copper complex, the peaks are broader due to the charging of the sample during data collection, affecting particularly the carbon species. Consequently, it is possible to decompose the C1s spectra of Cu-Lactose in more than three peaks (Fig. 3.13c), with the extra peak centred at 285.84 eV.

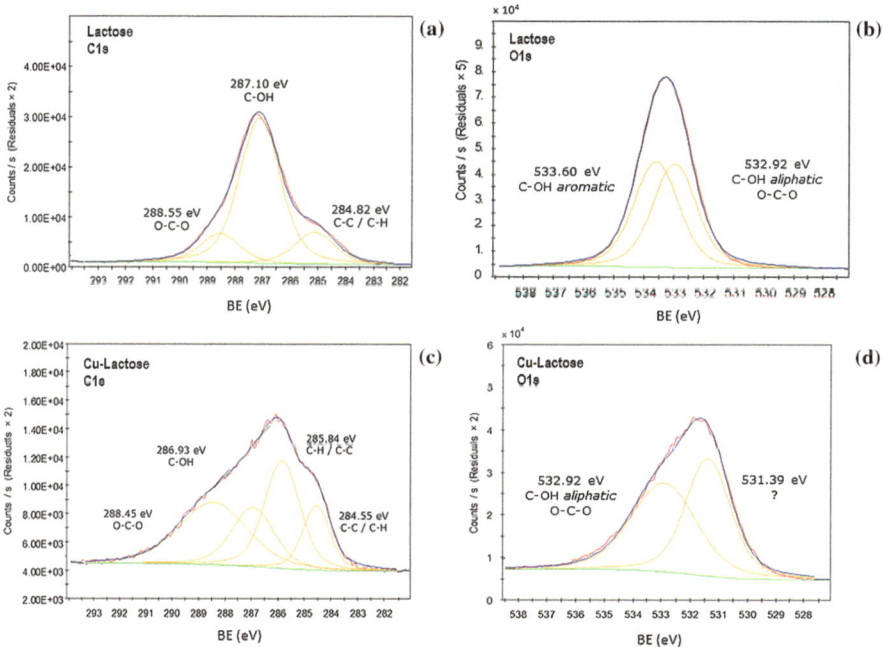

Fig. 3.13 Spectra of C1s and O1s regions of Lactose (**a–b**) and Cu-Lactose (**c-d**)

The O1s spectra for lactose and Cu-Lactose also differ. Complexation with copper resulted in the disappearance of the C–OH aromatic peak at 533.60 eV and the appearance of a new peak centred at 531.39 eV (Fig. 3.13d). Although dehydroxylation of saccharides can occur during XPS analysis [32], there was no sign of degradation of the Cu-Lactose sample during or after analysis. Therefore, the peak at 531.39 eV in the Cu-Lactose O1s spectra could be attributed to Cu–O bonds, which in inorganic compounds appear between 530.99–531.25 eV [14].

3.2.4 XRD

The diffractogram obtained for Cu-Lactose (not shown) was practically continuous. The few peaks present could be assigned to a cuprite (Cu_2O) phase in a largely amorphous sample.

3.2.5 Mass Spectrometry

3.2.5.1 Direct Mass Measurement by FIA

No ion corresponding to Cu-Lactose was found in any of the solvents. Lactose, albeit at low ion counts, was identified in all solvents. Its highest relative ion count was achieved in acetonitrile, where ions corresponding to the copper-acetonitrile complex were also identified.

3.2.5.2 GC-QTOF-MS with TSP

Lactose converts into its monosaccharide units after long exposure to temperatures above 180 °C [33], yielding characteristic anhydro sugars and furan derivatives at 650 °C [34]. The high number of chromatographic peaks—each representing at least one compound—obtained for Cu-Lactose in EI and CI modes (Fig. 3.14) was, therefore, expected as the thermal separation probe (set at 320 °C) acts as a low-temperature pyrolysis chamber. The identification of levoglucosenone (peak 4 in Fig. 3.14) is therefore significant, as this compound is a derivative of levoglucosan and a pyrolysis product of glucose [35] and lactose [36].

The EI chromatogram for Cu-Lactose contain bigger peaks corresponding to fatty acids (4 and 5 in Fig. 3.14a) than those found in the lactose EI chromatogram (not shown). As the same amount of lactose and Cu-Lactose were analysed, this result may indicate copper's preference for forming complexes with carboxylic acids originally present in lactose as contaminants. A strong 61 *m/z* peak, (Fig. 3.14b) and peak 3 (identified as 5-acetoxynethyl-2-furaldehyde) were also not observed for lactose. Their presence may be explained by the catalytic effect of Cu(II) ions, which

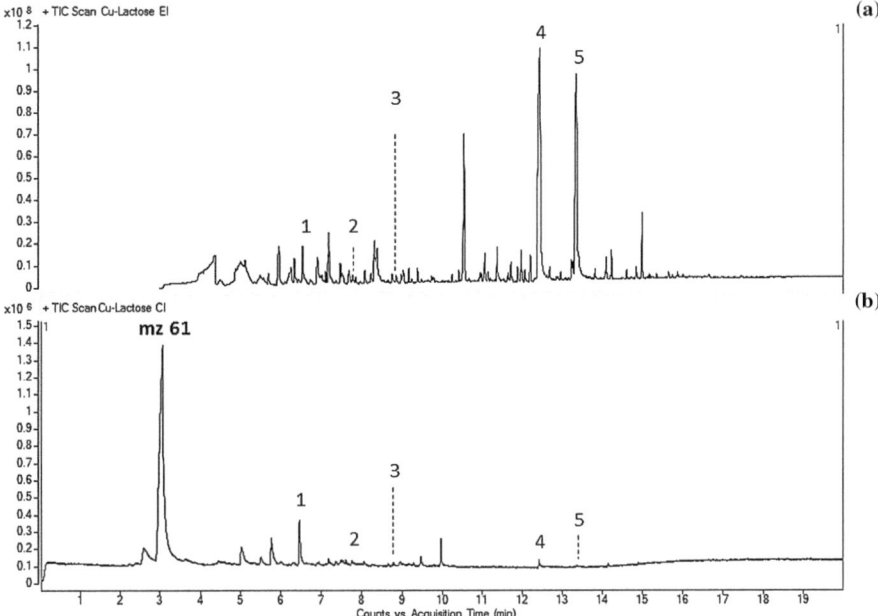

Fig. 3.14 EI (**a**) and CI (**b**) chromatograms for Cu-Lactose [1] 2H-Pyran-2,6(3H)-dione; [2] levoglucosenone; [3] 5-acetoxymethyl-2-furaldehyde; [4] n-hexadecanoic acid and [5] octadecanoic acid

lowers the decomposition temperature of carbohydrates. For cellulose, this effect has translated into higher yields of acetic acid, furans and gases [37].

3.2.6 Discussion

The characterization of Cu-Lactose by FTIR/Raman was complicated by poor peak definition and fluorescence. The XPS data confirmed the presence of Cu(II) ions within the compound suggesting coordination via the deprotonation of hydroxyl groups of the saccharide molecule. The identification of a cuprite phase by XRD is not surprising given that the glucose moiety in lactose is a reducing agent via its free aldehyde function that easily oxidizes to acid [20].

The mass spectrometry characterization of the Cu-Lactose was only possible using the GC-QTOF-MS with TSP method via identification of characteristic carbohydrate thermal decomposition products.

3.3 Characterization of Cu-Casein

The (+) copper coupon immersed in the electrolyte containing casein was covered by a translucent blue/greenish film at the end of the experiment (Fig. 3.1f), which peeled off upon drying out. From this point onwards, this film will be referred to as Cu-Casein.

The structure of a protein is organized in four different levels. Its primary structure is determined by the amino acid composition of the main polypeptide chain and is dominated by covalent bonds between atoms. The folding of the polypeptide chain results in different types of interaction which are responsible for their secondary, tertiary and quaternary structures.

Bovine casein is a mixture of αs1-, αs2-, β-, κ-casein and other proteinaceous impurities. Despite having many hydrophobic amino acids, caseins have an open hydrated structure with areas of high negative charge enabling them to form colloidal aggregates [38].

Complexation between metal ions and proteins requires coordination with a donor atom, possibly via the deprotonation of a nitrogen from an amino group, the oxygen from a carboxylic acid group and/or by interactions with the π-electrons from aromatic rings in the side chain of amino acids. In milk, αs- and β-caseins bind Ca^{2+} ions at phosphorylated serine residues [39] providing stability to micelle structures [40].

3.3.1 FTIR

The spectra (not shown) for casein and Cu-Casein were practically identical. The only minor differences between the FTIR bands of casein and Cu-Casein (Table 3.10) are the sharpening of the 3700–3300 cm^{-1} broad band, a decrease in the amide I band from 1651 to 1647 cm^{-1} and an increase in the amide II band from 1536 to 1540 cm^{-1}.

3.3.2 Raman

The Raman spectrum of casein is arched due to fluorescence (Fig. 3.15a). Fluorescence in proteins results from the movement of electrons in conjugated systems, such as those present in double bonds of aromatic rings. Changes to these systems lead to fluorescence suppression as observed for Cu-Casein (Fig. 3.15b).

Despite the issues with fluorescence, a few band assignments were still possible (Table 3.11).

Table 3.10 Characteristic FTIR bands in Casein and Cu-Casein

Casein		Cu-Casein		Band Assignments
3700–3300	br	3700–3300	br	N–H and O–H stretching vibrations
3072	sh	3071	sh	Overtone of amide II band
2961	m	2961	m	NH_{3+} and CH stretching vibrations
2870	vw	2877	w	CH stretching vibrations
1651	vs	1647	vs	Amide I band α helix (80% C=O stretching or COO^- asymmetric stretching vibrations; 10% C-N stretching vibrations and 10% N–H vibrations)
1536	s	1540	s	Amide II band and C=O stretching or COO^- asymmetric stretching vibrations
1335	vw	1335	vw	C–H deformation vibrations
1238	m	1238	m	Amide III α helix
1073	m	1070	m	C–C stretching vibrations

br (broad), sh (shoulder), vs (very strong), s (strong), m (medium), w (weak)

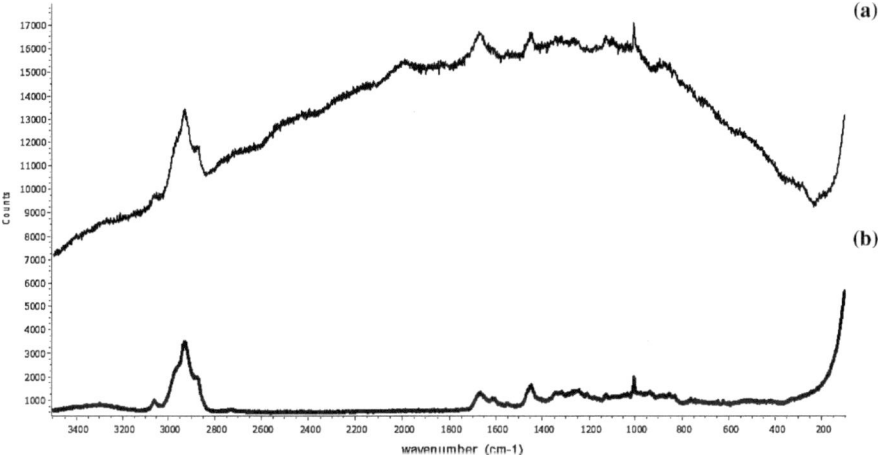

Fig. 3.15 Raman spectra of Casein (**a**) and Cu-Casein (**b**)

The results obtained from FTIR and Raman were predictable given how challenging metalloproteins are to analyse by these techniques, often requiring deconvolution and specific excitation to enhance the vibration of certain bonds. In the case of blue proteins—those that contain copper(II) ions—it is possible to observe shifts due to Cu–N, Cu–S or Cu–O bonds using a 600 nm laser [3]. The Raman data shown in Table 3.11 was generated with a 514 nm laser.

Table 3.11 Characteristic Raman bands of Casein and Cu-Casein

Casein		Cu-Casein		Band assignments
3061	w	3061	w	Overtone of amide II band
2934	vs	2934	m	Asymmetric CH$_2$ stretching
2879	sh	2879	sh	Symmetric CH$_2$ stretching
1673	s	1670	s	Amide I band (random chain) and COO$^-$ asymmetric stretching vibration
1555	w	1555	w	Amide II band
1323	vw			Amide III band (incl. CN stretching, NH bending, C–O stretching, O=C–N bending, etc.)
1211	w	1211	w	C–O vibrations
1073	m	1070	m	C–C stretching vibrations

sh (shoulder), vs (very strong), s (strong), m (medium), w (weak), vw (very weak)

3.3.3   XPS

The relative atomic percentages obtained for casein were 75.0% carbon, 16.96% oxygen and 8.03% nitrogen. Those for Cu-Casein were 72.2% carbon, 16.34% oxygen, 9.36% nitrogen, 0.6% phosphorus and 1.5% copper. The satellite peaks in the Cu2p spectrum of Cu-Casein indicate the presence of Cu(II) ions, assumed to be present given the characteristic blue colour of Cu-Casein.

The decomposed peaks for C1s and O1s and for N1s in casein and Cu-Casein are also informative (Fig. 3.16). Peak decomposition and assignment are complicated by their broad shape due to sample charging and the structural complexity of casein,

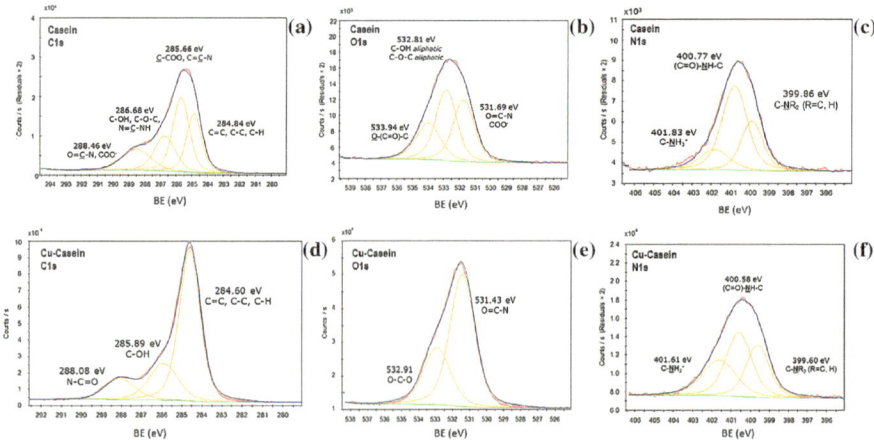

Fig. 3.16 Spectra of C1s, O1s and N1s regions of Casein (**a–c**) and Cu-Casein (**d–f**) peak fittings based on [32, 45]

which contains a variety of chemical environments. Notwithstanding these difficulties, there are clear differences between casein and Cu-Casein. The peaks of the C1s and O1s regions in Cu-Casein (Fig. 3.16d, e) are narrower than casein's (Fig. 3.16a, b), with the opposite effect observed for the N1s region (Fig. 3.16c, f).

It is tempting to interpret an increased oxygen carboxylate peak at 531.69 eV and, due to Brønsted donation [41], an increase in the nitrogen protonated amide peak at 401.61 eV, as evidence of Cu(II) coordination with the second carboxylic acid groups of aspartic and glutamic acids in the peptide chain. These findings contrast with values observed in biological systems, where copper coordination is predominantly via nitrogen and sulfur donor atoms in the side chains of histidine, cysteine and methionine [42] and, at pH values above 10, with the amine of lysine [43]. The sensitivity of XPS compounded by the structural complexity of proteins means that such deductions would have to be confirmed by supplementary analysis such as magnetic nuclear resonance (NMR) or electron paramagnetic resonance (EPR) [44].

3.3.4 XRD

The diffractogram obtained for Cu-Casein was similar to that obtained for Cu-Lactose. It was characteristic of an amorphous substance, with the exception of a peak above 35° 2θ, identified as a Cu_2O phase.

3.3.5 Mass Spectrometry

3.3.5.1 Direct Mass Measurement by FIA

No high mass ions were detected in any of the solvents, indicating that Cu-Casein is not soluble in acetonitrile, chloroform or methanol.

3.3.5.2 GC-QTOF-MS with TSP

The peaks in the chromatograms for casein and Cu-Casein in EI mode (not shown) relate to nitrogen containing aromatic compounds, products of the cyclisation of amino acids in the polypeptide chain subjected to pyrolysis [46].

Casein's chromatogram contained fewer peaks than Cu-Casein's. The extra peaks in the Cu-Casein chromatograms perhaps reflect specific thermal decomposition products associated with the presence of copper ions. Indole was the only compound identified in both EI and CI modes in Casein and Cu-Casein, which is one of the main products of the thermal degradation of bovine β-casein [47].

3.3.5.3 Proteomics

The four types of caseins were matched to the peptides recovered from casein and Cu-Casein (Table 3.12) but with different $-10\lg P$ values. The $-10\lg P$ value is a statistical indicator used to validate protein assignments (20 being the threshold for publication). Its values for the protein matches for casein and Cu-Casein are well above this minimal threshold, thus, validating the extraction and digestion protocol. The lower number of peptides identified in Cu-Casein may reflect metal complexation. Although metal chelation by proteins is extremely complex [48] and dependent on many factors [49], some tentative observations may be made by comparing the peptide coverage of casein and the Cu-Casein (Fig. 3.17).

For αs1-casein (Fig. 3.17a), half of the unidentified peptide residues in Cu-Casein contain tryptophan (W). Part of a group of amino acids that includes phenylalanine (F), tyrosine (Y) and histidine (H), tryptophan is reported to form non-covalent bonds with Cu(II) ions via the π-electrons from its aromatic ring [50, 51]. The other half of the unidentified peptide fragments contain histidine, an essential metal chelating site [52]. Metal chelation with histidine is via its imidazole ring supported by the amino group [53] or a deprotonated amide nitrogen supported by a carboxylate oxygen [54].

For αs2-casein (Fig. 3.17b), the unidentified peptide residues in Cu-Casein imply an interaction between histidine and glutamic acid (E) which could interact with Cu(II) ions via its deprotonated carboxylic groups. Metal complexation with the peptide chain of αs2-casein may also have been facilitated by its low hydrophobicity in relation to the other caseins.

For β-casein (Fig. 3.17c), the unidentified peptide residues in the copper complex appear to indicate an interaction with glutamic acid alongside serine (S), with tryptophan and with proline (P). Proline is the only amino acid that contains a second amide, so Cu–N coordination is not possible [55]. Consequently, coordination with copper only happens when proline is in the N-terminal position of the peptide chain. In other positions, proline residues force the peptide chain to bend [56], thus, facilitating the formation of macro-chelating sites [57].

An important caveat of the peptide coverage discussion is that the data is restricted to the most intense peptide fragments. Only further experiments and characterization with techniques such as NMR could elucidate metal-peptide chelation sites.

Table 3.12 Protein matches for Casein and Cu-Casein

Protein	Average Mass (Da)	Coverage (%)		$-10\lg P$*		# Peptides (# unique)	
		Casein	Cu-Casein	Casein	Cu-Casein	Casein	Cu-Casein
αs1-casein	24,529	75	59	259.41	191.20	70 (68)	19 (19)
αs2-casein	26,019	64	39	217.78	130.88	47 (47)	12 (12)
β-casein	25,107	91	73	208.26	190.11	49 (49)	28 (28)
κ-casein	21,269	71	43	247.12	194.55	64 (63)	18 (18)

*The $-10\lg P$ value is a statistical indicator used to validate protein assignments (20 being the threshold for publication)

Fig. 3.17 Peptide chain for αs1-2-caseins (**a, b**) β-casein (**c**) and k-casein (**d**) with identified peptides in Casein (underlined) and Cu-Casein (in bold) with short peptide sequences containing tryptophan (W) within a dashed square. A list of amino acids abbreviations is included in the Appendix

3.3.6 Discussion

Evidence of copper complexation with casein has been obtained with spectroscopic and mass spectrometry techniques. Notwithstanding the similarities between the FTIR spectra of casein and Cu-Casein, the slight changes in the bands of the latter indicate an interaction between copper and nitrogen/oxygen atoms. In the Raman spectra, the most obvious effect of copper complexation was the quenching of fluorescence, which may indicate coordination with amino acids containing aromatic rings.

Fluorescence in proteins usually arises from the amino acid residues phenylalanine, tyrosine and tryptophan, the latter a popular marker for studying proteins' interaction with cofactors [58–63]. These amino acids are present in casein [64] and are known to coordinate with Cu(II) ions [65–67].

The presence of Cu(II) ions in Cu-Casein, anticipated given its blue colour, was confirmed by XPS. These ions appear to catalyse the thermal degradation of casein given that the chromatogram of the Cu-Casein obtained by GC-QTOF-MS with TSP contained more peaks than that of pure casein. Nonetheless, indole—one of the main pyrolysis product of casein—was identified in casein and Cu-Casein.

The peptide fragments obtained from Cu-Casein matched all four types of caseins but at a lower coverage and intensity than those from pure casein sample. The difference in peptide coverage appears to correlate with sites of coordination when interpreted alongside the FTIR and Raman spectra. The data indicate different centres of interactions between the Cu(II) ions and deprotonated nitrogen, oxygen and amino acids containing aromatic rings as reported in the literature (for reviews on metal ion complexes, see, for example, [68, 69]) and as observed in blue copper proteins [42, 70, 71].

References

1. Socrates G (2001) Infrared and Raman characteristic group frequencies. Wiley & Sons, Chichester
2. Satake I, Matuura R (1961) Studies with copper (II) soaps: Part I. structural investigations of copper soaps and their complexes with pyridine and dioxane in solid state. Colloid Polym Sci 176(1):31–38
3. Nakamoto K (1997) Infrared and Raman spectra of inorganic and coordination compounds. Part B: Applications in coordination, organometallics and bioinorganic chemistry. Wiley & Sons Inc, New York/Chichester
4. Gunn M, Chottard G, Rivière E et al (2002) Chemical reactions between copper pigments and oleoresinous media. Stud Conserv 47(1):12–23
5. Prati S, Bonacini I, Sciutto G et al (2016) ATR-FTIR microscopy in mapping mode for the study of verdigris and its secondary products. Appl Phys A 122(1):1–16
6. La Nasa J, Lluveras-Tenorio A, Modugno F et al (2018) Two-step analytical procedure for the characterization and quantification of metal soaps and resinates in paint samples. Herit Sci 6(1):1–10

7. Salvadó N, Buti S, Pradell T et al (2019) Identification and distribution of metal soaps and oxalates in oil and tempera paint layers in fifteenth-century altarpieces using synchrotron radiation techniques. In: Casadio F, Keune K, Noble P et al (eds) Metal soaps in art—conservation and research. Springer, Cham, pp 195–210

8. Mathey Y, Greig DR, Shriver DF (1982) Variable-temperature Raman and infrared spectra of the copper acetate dimer $Cu_2(O_2CCH_3)_4(H_2O)_2$ and its derivatives. Inorg Chem 21(9):3409–3413

9. Conti C, Striova J, Aliatis I et al (2014) The detection of copper resinate pigment in works of art: contribution from Raman spectroscopy: detection of copper resinate pigment in works of art. J Raman Spectrosc 45(11–12):1186–1196

10. Doyle A, Felcman J, Gambardella MTdP, Verani CN, et al (2000) Anhydrous copper(II) hexanoate from cuprous and cupric oxides. The crystal and molecular structure of $Cu_2(O_2CC5H11)4$. Polyhedron 19(26–27):2621–2627

11. Robinet L, Corbeil M-C (2003) The characterization of metal soaps. Stud Conserv 48(1):23–40

12. Frost DC, Ishitani A, McDowell CA (1972) X-ray photoelectron spectroscopy of copper compounds. Mol Phys 24(4):861–877

13. Larson PE (1974) X-ray induced photoelectron and auger spectra of Cu, CuO, Cu_2O, and Cu_2S thin films. J Electron Spectros Relat Phenomena 4(3):213–218

14. Biesinger MC, Lau LWM, Gerson AR et al (2010) Resolving surface chemical states in XPS analysis of first row transition metals, oxides and hydroxides: Sc, Ti, V, Cu and Zn. Appl Surf Sci 257(3):887–898

15. Cano E, Torres CL, Bastidas JM (2001) An XPS study of copper corrosion originated by formic acid vapour at 40% and 80% relative humidity. Mater Corros 52(9):667–676

16. Ávila-Torres Y, Huerta L, Barba-Behrens N (2013) XPS-characterization of heterometallic coordination compounds with optically active ligands. J Chem 2013:1–9

17. Corbeil M-C, Robinet L (2002) X-ray powder diffraction data for selected metal soaps. Powder Diffr 17(1):52–60

18. Corkery RW (1997) Langmuir−Blodgett (L−B) multilayer films. Langmuir 13(14):3591–3594

19. Kolthoff IM, Coetzee JF (1957) Polarography in acetonitrile. II. Metal ions which have significantly different polarographic properties in acetonitrile and in water. Anodic waves. Voltammetry at rotated platinum electrode. JACS 79(8):1852–1858

20. Morriosn RT, Boyd RN (1992) Organic chemistry. Prentice-Hall International, London

21. Asomaning J, Mussone P, Bressler DC (2014) Pyrolysis of polyunsaturated fatty acids. Fuel Process Technol 120:89–95

22. Sembiring KC, Minami E, Kawamoto H et al (2020) Oxidative cleavage of linoleic and linolenic acids followed by decarboxylation for hydrocarbon production. J Jpn Inst Energy 99:1–7

23. Baer DR, Engelhard MH, Lea AS (2003) Introduction to surface science spectra data on electron and X-ray damage: sample degradation during XPS and AES measurements. Surf Sci Spectra 10(1):47–56

24. La Nasa J, Modugno F, Aloisi M et al (2018) Development of a GC/MS method for the qualitative and quantitative analysis of mixtures of free fatty acids and metal soaps in paint samples. Anal Chim Acta 1001:51–58

25. Gottschaldt M, Wegner R, Görls H et al (2004) Binuclear copper(II) complexes of 5-N-(β-ketoen)amino-5-deoxy-1,2-O-isopropylidene-α-D-glucofuranoses: synthesis, structure, and catecholoxidase activity. Carbohydr Res 339(11):1941–1952

26. Cerchiaro G, Sant'Ana AC, Temperini MLA, et al (2005) Investigations of different carbohydrate anomers in copper(II) complexes with d-glucose, d-fructose, and d-galactose by Raman and EPR spectroscopy. Carbohydr Res340(15):2352–2359

27. Norris KP, Greenstreet JES (1958) Infra-red absorption spectra of casein and lactose. Nature 181(4604):265–266

28. Listiohadi Y, Hourigan JA, Sleigh RW et al (2009) Thermal analysis of amorphous lactose and α-lactose monohydrate. Dairy Sci Technol 89(1):43–67

29. Bandwar RP, Srinivasa Raghavan MS, Rao CP (1995) Transition metal-saccharide chemistry: d-glucose complexes of Mn(II), Co(II), Ni(II), Cu(II) and Zn(II). Biometals 8(1)

30. Cebeci-Maltaş D, Alam MA, Wang P et al (2017) Photobleaching profile of Raman peaks and fluorescence background. Eur Pharm Rev 22(6):18–21

31. Kagan MR, McCreery RL (1994) Reduction of fluorescence interference in raman spectroscopy via analyte adsorption on graphitic carbon. Anal Chem 66(23):4159–4165

32. Stevens JS, Luca AC, Pelendritis M et al (2013) Quantitative analysis of complex amino acids and RGD peptides by X-ray photoelectron spectroscopy (XPS). Surf Interface Anal 45(8):1238–1246

33. Paez M, Martinez-Castro I, Olano A (1987) Thermal degradation of different crystalline forms of lactose. J Anal Appl Pyrolysis 12(1):31–38

34. da Silva PM, Gauche C, Gonzaga LV et al (2016) Honey: chemical composition, stability and authenticity. Food Chem 196:309–323

35. Pilath HM, Nimlos MR, Mittal A et al (2010) Glucose reversion reaction kinetics. J Agric Food Chem 58(10):6131–6140

36. Torri C, Lesci IG, Fabbri D (2009) Analytical study on the production of a hydroxylactone from catalytic pyrolysis of carbohydrates with nanopowder aluminium titanate. J Anal Appl Pyrolysis 84(1):25–30

37. Terakado O, Amano A, Hirasawa M (2009) Explosive degradation of woody biomass under the presence of metal nitrates. J Anal Appl Pyrolysis 85(1):231–236

38. Alaimo MH, Farrell HM Jr, Germann MW (1999) Protein structure and molecular enzymology. BBA 1431(2):410–420

39. Dalgleish DG, Parker TG (1980) Binding of calcium ions to bovine αsl-casein and precipitability of the protein–calcium ion complexes. J Dairy Res 47(1):113–212

40. Dalgleish DG (2011) On the structural models of bovine casein micelles—review and possible improvements. Soft Matter 7:2265–3227

41. Eggleston DS, Feldman SH (1990) Structure of the fibrinogen binding sequence: arginylglycylaspartic acid (RGD). Int J Pept Protein Res 36(2):161–166

42. Rubino T, Franz KJ (2012) Coordination chemistry of copper proteins: how nature handles a toxic cargo for essential function. J Inorg Biochem 107(1):129–143

43. Jeżowska-Bojczuk M, Stokowa-Sołtys K (2018) Peptides having antimicrobial activity and their complexes with transition metal ions. Eur J Med Chem 143:997–1009

44. Stevenson MJ, Janisse SE, Tao L et al (2020) Elucidation of a copper binding site in proinsulin C-peptide and its implications for metal-modulated activity. Inorg Chem 59(13):9339–9349

45. Rouxhet PG, Genet MJ (2011) XPS analysis of bio-organic systems. Surf Interface Anal 43(12):1453–1470

46. Moldoveanu Ş (1998) Analytical pyrolysis of natural organic polymers. Elsevier, Amsterdam/New York

47. Orsini S, Parlanti F, Bonaduce I (2017) Analytical pyrolysis of proteins in samples from artistic and archaeological objects. J Anal Appl Pyrolysis 124:643–657

48. Kozlowski H, Potocki S, Remelli M et al (2013) Specific metal ion binding sites in unstructured regions of proteins. Coord Chem Rev 257(19–20):2625–2638

49. Dudev T, Lim C (2014) Competition among metal ions for protein binding sites: determinants of metal ion selectivity in proteins. Chem Rev 114(1):538–556

50. Remko M, Fitz D, Broer R, Rode BM (2011) Effect of metal Ions (Ni^{2+}, Cu^{2+} and Zn^{2+}) and water coordination on the structure of L-phenylalanine, L-tyrosine, L-tryptophan and their zwitterionic forms. J Mol Model 17:3117–3128

51. Kadej A, Kuczer M, Czarniewska E et al (2016) High stability and biological activity of the copper(II) complexes of alloferon 1 analogues containing tryptophan. J Inorg Biochem 163:147–161

52. Rulíšek L, Vondrášek J (1998) Coordination geometries of selected transition metal ions (Co^{2+}, Ni^{2+}, Cu^{2+}, Zn^{2+}, Cd^{2+}, and Hg^{2+}) in metalloproteins. J Inorg Biochem 71(3–4):115–127

53. Altun Y, Köseoĝlu F (2005) Stability of Copper(II), Nickel(II) and Zinc(II) binary and ternary complexes of histidine, histamine and glycine in aqueous solution. J Solution Chem 34:213–231

54. Dunbar RC, Martens J, Berden G et al (2018) Binding of divalent metal ions with deprotonated peptides: do gasphase anions parallel the condensed phase? J Phys Chem A 122(25):5589–5596

55. Siemion IZ, Kubik A, Jezowska-Bojczuk M et al (1984) The absolute configuration on the chiral nitrogen atom of proline residue in the metal complexes of oligopeptides. J Inorg Biochem 22(2):137–141
56. Krieger F, Möglich A, Kiefhaber T (2005) Effect of proline and glycine residues on dynamics and barriers of loop formation in polypeptide chains. JACS 127(10):3346–3352
57. Pettit LD, Steel I, Formicka-Kozlowska G et al (1985) The L-proline residue as a 'break-point' in metal–peptide systems. Dalton Trans 3:535–539
58. Bent DV, Hayon E (1975) Excited state chemistry of aromatic amino acids and related peptides. III. Tryptophan. JACS 97(10):2612–2619
59. Chen Y, Barkley MD (1998) Toward understanding tryptophan fluorescence in proteins. Biochem 37(28):9976–9982
60. Vivian JT, Callis PR (2001) Mechanisms of tryptophan fluorescence shifts in proteins. Biophys J 80(5):2093–2109
61. Tayeh N, Rungassamy T, Albani HR (2009) Fluorescence spectral resolution of tryptophan residues in bovine and human serum albumins. J Pharm Biomed Anal 50(2):107–116
62. Li YH, Wang WJ, Xu XJ et al (2015) Changes in fluorescence intensity induced by soybean soluble polysaccharide–milk protein interactions during acidification. J Dairy Sci 98(12):8577–8580
63. Lai CW, Schwab M, Hill SC et al (2016) Raman scattering and red fluorescence in the photochemical transformation of dry tryptophan particles. Opt Express 24(11):11654–11667
64. Gordon WG, Semmett WF, Cable RS et al (1949) Amino acid composition of α-casein and β-casein. J Am Chem Soc 71(10):3293–3297
65. Davies RR, Kuang H, Qi D et al (1999) Artificial metalloenzymes based on protein cavities: exploring the effect of altering the metal ligand attachment position by site directed mutagenesis. Bioorg Med Chem Lett 9(1):79–84
66. Reddy PR, Manjula P (2007) Mixed-ligand copper(II)-phenanthroline-dipeptide complexes: synthesis, characterization, and DNA-cleavage properties. Chem Biodivers 4(3):468–480
67. Husain A, Kumar G, Sood T et al (2018) Synthesis, structural characterization and DFT analysis of an unusual tryptophan copper(II) complex bound via carboxylate monodentate coordination: Tetraaquabis(l-tryptophan) copper(II) picrate. Inorganica Chim Acta 482:324–332
68. Bal W, Sokołowska M, Kurowska E et al (2013) Binding of transition metal ions to albumin: sites, affinities and rates. BBA 1830:5444–5455
69. Sovago I, Varnagy K, Lihi N et al (2016) Coordinating properties of peptides containing histidyl residues. Coord Chem Rev 327–328:43–54
70. Nakamura K, Go N (2005) Function and molecular evolution of multicopper blue proteins. CMLS 62(18):2050–2066
71. Pérez-Henarejos SA, Alcaraz LA, Donaire A (2015) Blue copper proteins: a rigid machine for efficient electron transfer, a flexible device for metal uptake. Arch Biochem Biophys 584:134–148

Chapter 4
Characterization of "Cu-Milk Corrosion"

To illustrate the application of the data obtained from the characterization of copper-organic complexes presented in Chap. 3, another electrochemical experiment was conducted: the corrosion of copper coupons in bovine milk.

Bovine milk was chosen because cattle domestication was widespread in the ancient world [1–5] and milk was used as a paint binder [6–8].

For this experiment, the copper coupons were immersed in full-fat bovine milk and the power supply set to maintain 0.01A minimum current through the system over a period of two days. At the end of the experiment, the (+) coupon was covered in a bluish-green waxy deposit. Both coupons were carefully removed, washed in deionized water and allowed to dry in a fume cupboard for seven days. Once dried, the Cu-Milk corrosion was removed from the coupon and stored in a glass vial. A sample of the milk used for this corrosion experiment was freeze-dried to create a solid control.

4.1 Fourier Transform Infrared Spectroscopy (FTIR)

The spectra obtained for Cu-Milk corrosion and freeze-dried milk contained poorly resolved peaks and were practically identical. The only differences appear below 1200 cm^{-1} (Table 4.1), where some of the band frequencies are similar to those obtained for lactose and the copper soaps. The broad peaks in the spectra overshadow any peak shifts that could indicate metal complexation sites.

© The Author(s), under exclusive license to Springer Nature Switzerland AG 2022
L. C. Carvalho, *Beyond Copper Soaps*,
SpringerBriefs in Applied Sciences and Technology,
https://doi.org/10.1007/978-3-030-97892-1_4

Table 4.1 Characteristic FTIR bands of freeze-dried milk and Cu-Milk corrosion

Freeze-dried milk		Cu-Milk corrosion		Band assignments[a]
3700–3300	br	3700–3300	br	N–H stretching vibrations
2924	s	2924	s	C–H asymmetric stretching
2856	s	2856	s	CH_2 symmetric stretching vibrations
1747	s	1747	s	(C=O) ester stretching vibrations
1718–1587	br	1715–1587	br	(C=O) carboxylic acids and Amide I band
1550	m	1550	m	Amide II band and CO_2^- stretching (for Cu-Milk corrosion)
1465	m	1465	m	CH_2 symmetric bending
1379	w	1379	w	CH_2 asymmetric bending
1311	vw	1311	vw	Amide III band, CH deformation
1243	w	1243	w	C–O vibrations
1165	m	1165	m	$[CH_2(C=O)OCH_2]$ asymmetric stretching + CH_2 symmetric bending (out of plane)
1132–928	br			C–C stretching (–C–O stretching; CH_2 bending)
		1100	m	CH_2 asymmetric bending (out of plane)
893	w			C–H out-of-plane ring stretching twisting (lactose)
		721	w	CO_2^- scissors vibrations (carboxylates)
		689–456	br	C–O out-of-plane deformation (carboxylates) at 688 cm^{-1} and O–C–O in-plane bending (galactose) at 587 cm^{-1}

[a] Based on FTIR tables and [9]
Broad (br), very strong (vs), s (strong), m (medium), w (weak), vw (very weak)

4.2 Raman Spectroscopy

The Raman spectrum of freeze-dried milk contains poorly defined bands (Table 4.2) and more fluorescence than that of Cu-Milk corrosion (Fig. 4.1). The fluorescence is possibly due to a high percentage of casein in the sample with poor band definition resulting from the variable particle sizes, which can adversely affect Raman scattering [10]. Notwithstanding these differences, both spectra contain characteristic bands belonging to milk proteins at 1611 and 1005 cm^{-1} and milk fat at 1743 cm^{-1}, with

Table 4.2 Characteristic Raman bands of freeze-dried milk and Cu-Milk corrosion

Freeze-dried milk		Cu-Milk corrosion		Band assignments[a]
		3400–3200	br	N–H and OH stretching vibrations
		3060	w	N–H stretching vibrations
2997–2830	br			H–CH$_2$ asymmetric stretching
		2932	vs	CH$_2$ asymmetric and symmetric stretching vibrations
		2854	vs	
1746	vw	1743	w	C=O stretching vibrations
		1611	w	C=O amide I and C=C stretching vibrations
1444	w	1454	m	CH$_2$ deformation
1304	w	1304	w	CH$_2$ torsion
1126	w	1126	w	(C–O) + (C–C) stretching and C–O–H deformation
		1005	m	Ring-breathing (phenylalanine)
854	m	854	w	C–C–H + C–O–C deformation (amino acids mode)
687	m			C–C–O
457	w	457	w	C–C–C deformation and C–O torsion
		123	vs	Cu–O

Based on [9, 10]

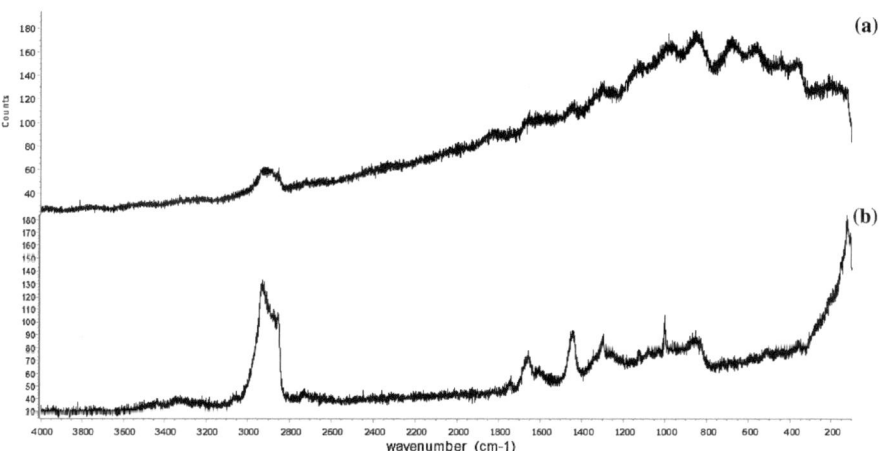

Fig. 4.1 Raman spectra of freeze-dried milk (**a**) and Cu-Milk corrosion (**b**)

Cu-Milk corrosion showing an increased fluorescence under 400 cm^{-1} and a very strong band at around 123 cm^{-1} that could be tentatively assigned to Cu–O.

4.3 X-ray Photoelectron Spectroscopy (XPS)

The XPS survey spectrum of Cu-Milk corrosion only revealed the presence of carbon, oxygen and copper. The absence of nitrogen from the results may indicate that compounds containing nitrogen such as proteins are located below 10 nm into the sample. The Cu2p spectrum contained satellite peaks confirming the presence of Cu(II) ions.

Some information can be obtained from the C1s and O1s regions (Fig. 4.2).

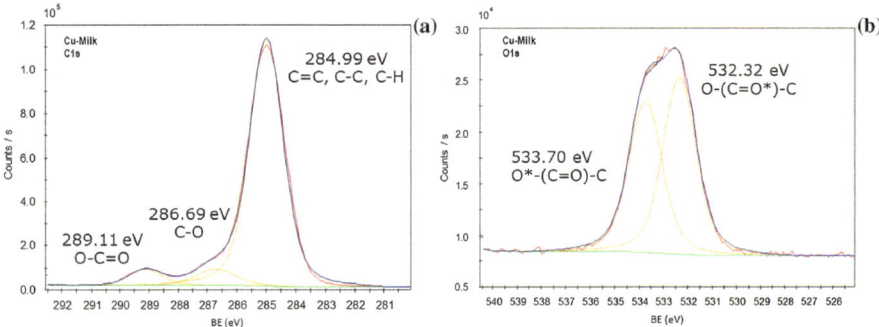

Fig. 4.2 Spectra of C1s (**a**) and O1s (**b**) regions of Cu-Milk corrosion

The carbon region of Cu-Milk corrosion (Fig. 4.2a) can be decomposed into three peaks. The first peak at 284.99 eV corresponds to the carbon chain. The peak at 286.69 eV for C–O species is within the energy range of C–OH species in lactose, Cu-Lactose and casein, whereas the peak for O–C=O at 289.11 eV species matches that of palmitic acid. For oxygen types (Fig. 4.2b), the peak at 533.70 eV for O*–(C=O)–C has similar values to the C–OH aromatics in lactose and casein.

4.4 Mass Spectrometry

4.4.1 Gas Chromatography with Quadrupole Time-of-Flight Mass Spectrometry using a Thermal Separation Probe (GC-QTOF-MS with TSP)

Some compounds identified in the freeze-dried milk and Cu-Milk corrosion have the same retention time as the corresponding ligands identified in the copper-organic complexes (Table 4.3). The differences between freeze-dried milk and Cu-Milk corrosion are restricted to carbohydrate derivatives. The presence of copper ions would have favoured the oxidation of alcohols (e.g. furan methanol) into aldehydes (e.g. 2-Furancarboxaldehyde, 5-methyl) and the formation of levoglucosenone.

Table 4.3 Identified compounds in freeze-dried milk and Cu-Milk corrosion cross-referenced with synthetized copper-organic complexes

RT (min)	Compound	Milk	Cu-milk	Complex
3.15–3.34	Formic acid	X	X	
3.72–3.75	Acetic acid	X		
4.86–4.96	3-Furanmethanol	X		
4.98–5.01	3-Furaldehyde	X	X	
5.36–5.41	2-Furanmethanol	X		
6.19–6.23	2-Furancarboxaldehyde, 5-methyl		X	
6.51–6.58	**2H-Pyran-2,6(3H)-dione**	X	X	Cu-Lactose
7.08–7.13	Furaneol	X		
7.46–7.52	**Levoglucosenone**		X	Cu-Lactose
8.85–8.92	**Indole**	X	X	Cu-Casein
9.08–9.16	n-Decanoic acid (C10:0)	X	X	
10.31–10.36	Dodecanoic acid (C12:0)	X	X	
11.42–11.49	Tetradecanoic acid (C14:0)	X	X	
12.43–12.54	**n-Hexadecanoic acid (C16:0)**	X	X	Cu-Palmitate
13.36–13.42	**Octadecanoic acid (C18:0)**	X	X	Cu-Stearate
14.50–14.64	Pyrrolo[1,2a]pyrazine-1,4-dione, hexahydro-3-(phenylmethyl)-	X	X	
18.05–18.21	Cholesterol	X	X	

[x] = present

Table 4.4 Identified compounds in the BSTFA-derivatized extracts of freeze-dried milk and Cu-Milk Corrosion cross-referenced with synthetized copper-organic complexes

RT (min)	Compound	Milk	Cu-Milk	Cu- complex
12.39	Nonanoic acid (C9:0)		X	
14.49	Dodecanoic acid (C12:0)	X		
16.87	Tetradecanoic acid (C14:0)	X		
18.01	n-pentadecanoic acid (C15:0)		X	
20.14	**n-Hexadecanoic acid (C16:0)**		X	Cu-palmitate
21.92	**Octadecanoic acid (C18:0)**		X	Cu-stearate
29.04	Cholesterol		X	

[x] = present

4.4.2 Solvent Extract Derivatized with N, O-Bistrifluoroacetamide (BSTFA) Analysed by Gas Chromatography with Mass Spectrometry (GC-MS)

Only two compounds—dodecanoic and tetradecanoic acid—were identified in freeze-dried milk extract compared to five in the Cu-milk corrosion extract (Table 4.4). The difference in composition between the extracts was probably due to protein interference. A few days after analysis, when the vials were removed from the equipment for disposal, the milk extract was opaque, whilst the Cu-Milk vials were still translucent.

4.4.3 Proteomics

In milk, the caseins are mostly arranged as micelles, in which k-casein concentrates on the surface, β-casein in the middle and the αs1-2-caseins distributed throughout, stabilized by Ca^{2+} ions [11]. As changes in pH, temperature, mineral and water content can affect the structure of casein micelles [12], the results obtained for freeze-dried milk may not be directly comparable to those obtained from Cu-Milk corrosion (Table 4.5).

Furthermore, although complexation between Cu^{2+} and the peptide chain may not be favoured at a neutral pH (the measured pH of the milk used for the experiment was 7), the copper ions could still have disrupted the casein micelles [13, 14], affected trypsin's enzymatic activity [15] and/or lead to mis-cleavage of the peptide chain during digestion [16]. Therefore, given these complexities, the identification of characteristic proteins in both samples with a method that did not include a metal chelation step [17] is a good result.

Table 4.5 Most intense proteins identified in freeze-dried milk and Cu-Milk corrosion

Protein	Coverage (%)		$-10\lg P^a$		# Peptides (# Unique)	
	Milk	Cu-Milk	Milk	Cu-Milk	Milk	Cu-Milk
αs1-casein	70	73	289.42	246.02	63 (62)	59 (59)
αs2-casein	61	57	223.61	167.00	40 (38)	27 (26)
β-casein	76	82	256.79	225.55	64 (64)	49 (49)
κ-casein	57	65	266.95	237.50	64 (63)	33 (32)
β-lactoglobulim	78	72	225.50	217.11	30 (29)	40 (39)

[a] The $-10\lg P$ value is a statistical indicator used to validate protein assignments (20 being the threshold for publication)

4.5 Discussion

The results obtained confirmed that the electrochemical experiment succeeded in creating a deposit that mimicked the characteristics of milk. Unfortunately, few insights into the compounds present in the sample could be obtained from FTIR or Raman, except for the presence of proteins.

Real insights into Cu-Milk corrosion were obtained from the mass spectrometry techniques. For lipids, the most diagnostic results were obtained using the thermal separation probe, enabling correlation with copper-organic complexes. Similarly, most chromatographic peaks remained unassigned as the NIST library search results could not be validated with analysis in chemical ionization mode.

The protein profiles obtained by proteomics characterized the freeze-dried milk and Cu-Milk corrosion samples as bovine milk, thus, illustrating the applicability of this technique to complex samples.

References

1. Copley MS, Berstan R, Dudd SN et al (2003) Direct chemical evidence for widespread dairying in prehistoric Britain. PNAS 100:1524–1529
2. Craig OE, Mulville J, Pearson MP et al (2000) Detecting milk proteins in ancient pots. Nature 408:312
3. Craig OE, Saul H, Lucquin A et al (2013) Earliest evidence for the use of pottery. Nature 496(7445):351–354
4. Dunne J, Evershed RP, Cramp LJE et al (2013) The beginnings of dairying as practised by pastoralists in 'Green' Saharan Africa in the 5th millennium BC. Doc Praehist 40:118–130
5. Grillo KM, Dunne J, Marshall F et al (2020) Molecular and isotopic evidence for milk, meat, and plants in prehistoric eastern African herder food systems. PNAS 117(18):9793–9799
6. American S (1847) Milk paint. Sci Am 2(37):296
7. Villa P, Pollarolo L, Degano I et al (2015) A milk and Ochre paint mixture used 49,000 years ago at Sibudu, South Africa. PLoS ONE 10(6):e0131273–e0131273
8. Colombini MP, Fuoco R, Giacomelli A et al (1998) Characterization of proteinaceous binders in wall painting samples by microwave-assisted acid hydrolysis and GC-MS determination of amino acids. Stud Conserv 43(1):33–41

9. Mendes TO, Junqueira GMA, Porto BLS et al (2016) Vibrational spectroscopy for milk fat quantification: line shape analysis of the Raman and infrared spectra: vibrational spectroscopy for milk fat quantification. J Raman Spectrosc 47(6):692–698

10. Mcgoverin CM, Clark ASS, Holroyd SE et al (2010) Raman spectroscopic quantification of milk powder constituents. Anal Chim Acta 673(1):26–32

11. Dalgleish DG, Corredig M (2012) The structure of the casein micelle of milk and its changes during processing. Annu Rev Food Sci Tec 3:449–467

12. Horne DS (2014) Casein micelle structure and stability. In: Singh H, Boland M, Thompson A (eds) Milk proteins: from expression to food. Academic Press, London

13. Sigel H, Bruce Martin R (1982) Coordinating properties of the amide bond. Stability and structure of metal ion complexes of peptides and related ligands. Chem Rev 82(4):385–426

14. Hellinga HW (1998) The construction of metal centers in proteins by rational design. Fold Des 3(1):R1–R8

15. Higaki JN, Haymore BL, Chen S et al (1990) Regulation of serine protease activity by an engineered metal switch. Biochem 29(37):8582–8586

16. Šlechtová T, Gilar M, Kalíková L et al (2015) Insight into trypsin miscleavage: comparison of kinetic constants of problematic peptide sequences. Anal Chem 87(15):7636–7643

17. Janecki DJ, Reilly JP (2005) Denaturation of metalloproteins with EDTA to facilitate enzymatic digestion and mass fingerprinting. RCM 19(10):1268–1272

Chapter 5
Conclusion

Traditionally, archaeological metal corrosion has been considered a mixture of inorganic compounds resulting from the oxidation of the metal surface by substances in the deposition environment. Although conservators routinely find macro-organic remains of historical importance preserved within corrosion products of composite objects, metal corrosion has seldom been considered a locus for the preservation of micro-organic remains such as food residues. Using electrochemical experiments, this research has investigated the preservation of organic molecules in copper corrosion via metal-organic complexation, testing different analytical techniques for their characterization.

The first observations is that copper(II) organic complexes have different textures to copper(II) inorganic compounds. Powdery residues were obtained for saturated carboxylic acids (palmitic and stearic) and sticky films for unsaturated acids (oleic and linoleic). Corroding copper in lactose resulted in a powdery residue, in a brittle translucent film with casein and a waxy residue with milk. This is significant as these characteristics can guide sampling of corrosion for organic residue studies, as currently this material is routinely discarded during the conservation of archaeological copper objects..

The information obtained by Fourier Transform Infrared (FTIR) analyses of copper-organic complexes decreased with the complexity of the organic ligand. Interpretation of a FTIR spectrum of a sample containing a copper-organic complex mixed with inorganic phases is challenging and dependant on the ratio of these compounds in the sample. This is an important result as FTIR is a popular technique for the identification of organic residues in museums.

Analysis by Raman spectroscopy corroborated the identification of coordination between copper and carboxylic acids but was less informative for the remaining complexes and the Cu-Milk corrosion due to fluorescence, that could not be circumvented. The usefulness of powder X-ray Diffraction (XRD) for the identification of copper-organic complexes was also limited by their mostly amorphous nature. This matters as unless copper-organic complexes are crystalline they cannot be identified

© The Author(s), under exclusive license to Springer Nature Switzerland AG 2022   49
L. C. Carvalho, *Beyond Copper Soaps*,
SpringerBriefs in Applied Sciences and Technology,
https://doi.org/10.1007/978-3-030-97892-1_5

by XRD, the most popular technique used for the characterization of archaeological metal corrosion.

The role of X-ray Photoelectron Spectroscopy (XPS) in this research was to confirm the state of oxidation of copper ions in the various copper-organic complexes. Although some tentative observations could also be made from the data obtained for the remaining elements in the samples, the findings would require validation by other techniques. The limitations of XPS were evident in the characterization of casein/Cu-Casein and Cu-Milk corrosion, which yielded broad peaks permitting alternative peak fitting options to those presented.

During mass spectrometry characterization, it was identified that the copper-organic complexes had a limited solubility in chloroform and methanol. This is meaningful as these are the most common solvents used for organic residue analysis.

Using a thermal separation probe (TSP) for the identification of lipids in Cu-Milk corrosion offered advantages over N, O-Bistrifluoroacetamide (BSTFA) derivatized solvent extract. Some laboratories already use TSP for screening archaeological samples for organic residues prior to further analysis with other protocols. But when only a small amount of sample is available for analysis and organic molecules coordinating with an inorganic matrix may be present, TSP is the only technique able to recover different classes of organic compounds within the same run and with minimum sample preparation. Its shortcomings are in the identification of proteins and temperature-sensitive compounds.

Further research is required to compare the performance of the bottom-up proteomics protocol used in this research with protocols containing a metal chelating step. This is because the use of metal chelating agents could have a negative impact on protein recovery yields from samples where proteins may be coordinating with metal ions, resulting in their elimination from the sample. Although laboratories currently overcome this effect through the analyses of samples above 200 mg, such a large sample requirement would exclude paintings and objects/residues under curatorial constraints.

Corrosion processes are extremely complex. To explain in more detail the various chemical reactions that led to the corrosion products produced by the electrochemical experiments described in this book would require additional experiments and data from other analytical techniques. Notwithstanding these constraints, electrolytic cells proved to be a fast and reliable way to test different corrosive environments and to produce copper-organic complexes at a relatively low cost.

In cultural heritage, and especially in archaeology, the identity of precursor substances of a residue is not known. Consequently, data from analytical techniques can only provide snapshots of the truth—never the whole truth. With a multi-analytical protocol it is possible to obtain a more holistic view of the sample in recognition that the boundaries between inorganic and organic chemistry are in fact artificial. It is hoped that the data presented in this book will open up different types of investigations and contribute to the effective allocation of resources.

Chapter 6
Analytical Techniques

6.1 Fourier Transform Infrared Spectroscopy (FTIR)[1]

Samples were analysed in reflection or in transmission mode.

In reflection mode, the technique used was Attenuated Total Reflectance (ATR) where around 1 mg of sample was pressed against the reflective crystal of the equipment or analysed using a microscope fitted with a micro-ATR objective pressed against the surface containing the sample (in this case the copper coupon).

ATR measurements were obtained with a Perkin Elmer Spotlight 200i FTIR Microscope System at the Heritage Science Laboratory at the Bodleian Library, University of Oxford.

In transmission mode, the technique used was the KBr pellet method. The sample was mixed with KBr (kept warm and desiccated before use) in an agate mortar with a pestle in a 2 mg/200 mg ratio. The mixture was transferred to a pellet die and pressure applied with a hydraulic pump to form a disc of around 1 mm thickness which was immediately transferred to the equipment for analysis. Transmittance measurements were taken with an Excalibur Series Varian UMA600 at Begbroke Science Park, University of Oxford.

Measurements were a combination of 64 scans between the 4000–400 cm^{-1} range and included background subtraction. Data was processed using Perkin Elmer Spectrum Software v.10.5.4 or Digilab Resolutions Pro 4.0 software and figures created with Spectragryph v.1.2.12. Band assignments were based on FTIR tables [1–3] and, where applicable, articles in the literature.

---

[1] For an introduction to FTIR, see [1].

L. C. Carvalho, *Beyond Copper Soaps*,
SpringerBriefs in Applied Sciences and Technology,
https://doi.org/10.1007/978-3-030-97892-1_6

6.2   Raman Spectroscopy[2]

Spectra were obtained using a Horiba™ Jobin Yvon spectrometer at Cranfield University operated by Dr. Chris Dyer. Samples were scraped off the copper coupon onto a glass slide. Each sample was run on a low laser power and short exposure to judge the level of Raman scattering and fluorescence and checked for visible sample damage.

Selected laser power and exposure time were those that maximized the Raman signal without causing damage to the sample. Typical values were 6 s under maximum power at 5-10 mW with excitation at 532 nm from a continuous-wave solid-state laser. The microscope system was based around a BX-41 microscope operating in epi-illumination mode (reflection), collecting Raman data on backscattered geometry. A × 50 metallurgical (dry) objective was used throughout except when a × 10 gave a longer distance between the end of lens and sample surface, at the expense of signal collection efficiency.

The spectrograph used to disperse the light was fitted with a cooled charge coupled device (CCD) to record the resulting signal. For practical purposes, there is an inverse relationship between reduction in laser power and increase in exposure time when using a CCD-based system: if the power is dropped ten-fold to avoid thermally affecting the sample, then the acquisition time is increased ten-fold to achieve a comparable S/N ratio. For fragile samples, 10% power with 120 s of exposure was used.

Data was processed and figures created with Spectragryph v.1.2.12. Band assignments were based on Raman Tables [2, 3] and, where applicable, articles in the literature.

6.3   X-Ray Photoelectron Spectroscopy (XPS)[3]

Spectra were obtained with Thermo Scientific™ K-Alpha™ X-ray photoelectron spectrometer at the Begbroke Science Park, University of Oxford operated by Dr. Phillip Holdway.

The system was calibrated for a carbon peak at 284.80 eV. A flood gun was used to stop the sample charging. The spot size was 400 μm. For a survey scan, a step size of 1 eV and dwell time of 10 ms were used and, for a detailed scan, a step size of 0.1 eV and dwell time of 50 ms. Two spots were measured for each sample, with peak fittings undertaken using Casa© software with assignments based on XPS reference spectra published by reference libraries and, where applicable, articles in the literature. Dr. Phillip Holdway performed curve fitting in consultation with the author.

---

[2] For an introduction to Raman, see [4].

[3] For an introduction to XPS, see [5].

6.4   X-Ray Diffraction (XRD)[4]

The archaeological samples were pulverized, mounted on a single silicon crystal plate and spectra obtained using a PANalytical X'Pert PRO Cu alpha instrument at the Crystallography Laboratory in the Department of Chemistry, University of Oxford. The equipment was set to operate in continuous mode at 40 kV/40 mA with a scanned area set between 1°–70° 2θ, 0.02 step size and 3° per minute.

The copper-organic complexes were analysed in a Bruker d* ADVANCE Eco X-ray Powder Diffractometer operating with a Cu tube at 40 kV 20 μA with a Lynxeye XE solid-state detector at Begbroke Science Park, University of Oxford, operated by Dr. Phil Holdway. The scanned area was set between 5°–40° 2θ, 0.02 step size and 1° per minute.

Data was processed using QualX© software with phase identification using the Crystallography Open Database (COD).

6.5   Mass Spectrometry Techniques[5]

Most analyses were performed at the Mass Spectrometry Facility at the Department of Chemistry, except 6.5.3, which was performed at the Research Laboratory for Archaeology and History of Art at the School of Archaeology, University of Oxford.

6.5.1   Direct Flow Injection Analysis (FIA)

For these analyses, around 0.35 mg of sample was placed in a glass vial and 50 μL of solvent (acetonitrile, chloroform or methanol) was added followed by 1 h sonication. After centrifugation, the supernatant was transferred to a 2 mL vial for injection into a Thermo Exactive High-Resolution Orbitrap FTMS without front end LC in direct infusion (loop injection) mode. Data analysis was performed using Thermo FreeStyle software.

---

[4] For an introduction to powder XRD, see [6].

[5] For the principles and application of mass spectrometry, see [7].

6.5.2   Gas Chromatography with Quadrupole Time-of-Flight Mass Spectrometry using a Thermal Separation Probe (GC-QTOF-MS with TSP)

A glass microvial containing 1 mg of pulverized sample was placed inside the TSP attached to an Agilent 7890B gas chromatograph equipped with a Restek Rxi-5 ms column (30 m × 320 μm × 0.25 μm). The mass spectrometer was an Agilent 7250 GC/Q-TOF equipped with a low-energy-capable EI source (70 eV) and an interchangeable CI source.

Different settings were used to optimize conditions so that analyses were relatively fast and reproducible. Heat-assisted derivatization with N, O-Bis(trimethylsilyl)trifluoroacetamide (BSTFA) was also attempted without any detectable change in the results. The best results were achieved with the TSP set at 300 °C and the oven temperature set at 40 °C for one minute, increasing by 20 °C/minute until it reached 320 °C where it was held for five minutes. Helium was used as a carrier gas, at 1.43 mL/min flow rate and 8.70 psi pressure. The equilibration time was set at 0.5 min and the sample injection was splitless. The mass range was 50–650 m/z. All samples were run in triplicate.

Data analysis was performed with Agilent Mass Hunter Qualitative Analysis 10.0 with compound assignments using NIST Library 17.

6.5.3   Proteomics

The proteomics method used was bottom-up proteomics, where protein matches are based on manually validated peptides using specialist software and protein databases.

Around 2 mg of sample was partially dissolved in 2:1v/v chloroform/methanol under sonication. After centrifugation, the supernatant was transferred to a clean tube and ice-cold acetone was added prior to overnight incubation at −20 °C to precipitate the proteins. The following day, after centrifugation, the supernatant was discarded and the precipitate extracted with RIPA Buffer (prepared in the laboratory). The protein extract was digested using LysC and trypsin followed by desalting and sample concentration using C18 ZipTip [8].

The tryptic digest was re-suspended in 40 μL milli-Q water with 2% acetonitrile and 0.1% formic acid. 2 μL was analysed by nanoLC-MS/MS using a Waters NanoAcquity-UPLC system interfaced with a Thermo LTQ Velos Orbitrap Elite mass spectrometer possessing an EASY-Spray ion source. Initial peptide trapping on a packed guard column (75 μm i.d. × 20 mm, Acclaim Pep-map100 C18, 3 μm, 120 Å) was carried out with solvent A (0.1% Formic Acid in water) at 140 bar. Peptides were separated on an EASY-spray Acclaim Pep-Map® analytical column (75 μm i.d. × 15 mm, RSLC C18, 3 μm, 100 Å) using a 120 min linear gradient ranging from 3 to 97% of solvent B (0.1% formic acid in acetonitrile), with 300 nL/min flow rate and 40 °C column temperature. The nanoESI source operated at 1600 V needle voltage with the temperature of the ion transfer tube set to 275 °C.

The separated peptides were ionized via electrospray directly into the mass spectrometer operating in a data-dependent acquisition mode using a CID-based method. Full scan MS spectra (scan range 350–1500 m/z, resolution 120,000, AGC target 1e6 and maximum injection time 250 ms) and subsequent CID MS/MS spectra (AGC target 5e4 and maximum injection time 100 ms) of the 10 most intense peaks were acquired in the Ion Trap. CID fragmentation was performed at 35% of normalized collision energy and the signal intensity threshold was kept at 500 counts.

Brand new guard and separation columns were used to avoid contamination from previous samples analysed in the laboratory. A Bovine Serum Albumin standard was also digested and analysed in parallel using the same conditions. Samples were run only once and with no duplicates.

Data analysis was performed using PEAKS 8.5 v (Bioinformatics Solutions Inc., Waterloo, ON, Canada). Raw MS data was searched against the general UniProt database and identifications were confirmed with UniProt taxonomy databases with LysCTryp selected as the protease. Carbamidomethylation (Cysteine) was set as a fixed modification and Oxidation (Methionine), Deamination (Asparagine and Glutamine) and +12 Da on N-terminal proline were set as variable modifications. Precursor mass tolerance was set as 15 ppm and fragment mass tolerance for CID was set to 0.8 Da.

PEAKS DB measures the quality of the Peptide Spectrum Match (PSM) internally with a Linear Discriminant Function (LDF) score. To identify the protein, LDF considers the matching of fragment ions and spectrum peaks and the similarity between the de novo sequencing peptide and the universal database peptide amongst other factors. The LDF score is converted to $-10\lg P$ to facilitate assessment. Only peptides with a $-10\lg P > 20$ (p-value of 1%) were selected for PSM validation with a Target Decoy PSM Validator node based on q-values at a 5% false discovery rate (FDR). Only validated peptides were used in UniProt protein database searches. Analyses and data processing were performed by Mrs. Elisabete Pires in consultation with the author.

6.5.4 Solvent Extract Derivatized with N, O-Bistrifluoroacetamide (BSTFA) Analysed by Gas Chromatography with Mass Spectrometry (GC–MS)

Around 2 mg sample was extracted with 2:1v/v chloroform/methanol mixture under sonication for 30 min. The extract was derivatized prior to GC–MS analyses by adding 20 μL of BSTFA with 5 μL tridecanoic solution (100 μg/μL in hexane) and kept for one hour at 70 °C. The solution was then transferred to glass vials containing micro-inserts and 10μL hexadecane solution (50 μg/g in iso-octane) added as an internal standard.

Analysis was performed using an Agilent 7820 A gas chromatograph equipped with a Restek Rxi-5 ms column (30 m × 0.25 mm i.d., 0.25 μm). The mass spectrometer was an Agilent 5975 quadrupole mass spectrometer, operating in EI mode (70 eV), and the scan range was m/z 50–650 amu. The inlet temperature 300 °C, flow rate 1.2 mL/min and transfer line temperature 280 °C were used with helium as the carrier gas.

The temperature program for the GC oven started with two minutes at 50 °C, followed by a ramp from 50 °C to 300 °C at 10 °C/min and finishing by 10 min isothermally held at 300 °C. Injections were made by an Agilent 7693 A autosampler and the sample injection volume was 1 μL in split mode.

Compounds were identified using the University of Pisa database of compounds.

References

1. Cross AD, Alan Jones R (1969) An introduction to practical infra-red spectroscopy. Butterworth & Co, London
2. Socrates G (2001) Infrared and Raman characteristic group frequencies. John Wiley & Sons, Chichester
3. Nakamoto K (1997) Infrared and raman spectra of inorganic and coordination compounds. Part B: applications in coordination, organometallics and bioinorganic chemistry. John Wiley & Sons Inc, New York
4. Gardiner DJ, Graves PR (1989) Practical Raman spectroscopy. Springer-Verlag, Heidelberg
5. van der Heide P (2011) X-ray photoelectron spectroscopy: an introduction to principles and practices. John Willey & Sons, Hoboken
6. Pecharsky V, Zavalij P (2009) Fundamentals of powder diffraction and structural characterization of materials. Springer, New York
7. Hoffmann ED, Stroobant V (2007) Mass spectrometry: principles and applications. Wiley, Chiscrester
8. Pires E, Carvalho LC, Shimada I et al (2021) Human blood and bird egg proteins identified in red paint covering a 1000-year-old gold mask from Peru. J Proteome Res 20(11):5212–5217

Appendix

Synthesis of copper palmitate[1]

1 g of palmitic acid dissolved in 50 mL of ethanol was mixed with 50 mL of a saturated solution of $CuSO_4$ in deionized water. The pH was adjusted to 9 with a saturated solution of NaOH in deionized water and the mixture was heated at 70 °C for 20 min. The mixture was filtered and the recovered blue precipitate was washed with warm ethanol (to remove the excess acid) followed by deionized water (to remove any $CuSO_4$) and allowed to dry at 100 °C for two hours.

Amino acids abbreviations

A: Alanine	N: Asparagine
C: Cysteine	O: Pyrrolysine
D: Aspartic acid	P: Proline
E: Glutamic acid	Q: Glutamine
F: Phenylalanine	R: Arginine
G: Glycine	S: Serine
H: Histidine	T: Threonine
I: Isoleucine	U: Selenocysteine
K: Lysine	V: Valine
L: Leucine	W: Tryptophan
M: Methionine	Y: Tyrosine

---

[1] Based on Corbeil M-C, Robinet L (2002) X-ray powder diffraction data for selected metal soaps. Powder Diffr 17(1):52–60.